Sal, Lumen & Spiritus Mundi Philosophici

or,

The dawning of the Day,
discovered by the Beams of Light:

Showing, the true Salt and Secret of the Philosophers, the first and universal Spirit of the World.

by Clovis Hesteau Nuisement

Facsimile of 1657 edition

Reprint 2012

Phoenix Press
Pocket Series

© Copyright 2012, Phoenix Press.
All rights reserved.
Edited by Joshua Arent.
ISBN: 978-1-300-16261-2

Sal, Lumen, & Spiritus Mundi Philosophici:

OR,

The dawning of the Day,
Discovered

By the Beams of Light:

Shewing,

The true SALT and SECRET of the Philosophers,

The first and universal Spirit of the World.

Written Originally in French, afterwards turned into Latin, By the Illustrious Doctor,

Lodovicus Combachius,

Ordinary Physitian to the King, and publick Professor of Physick in the University of *Mompelier*.

And now transplanted into *Albyons* Garden, By R. T. Φιλομαθ.

Printed at *London*, by *J. C.* for *Martha Harrison*, at the Lamb at the East-end of S. Pauls. 1657.

The Contents.

BOOK I.

Chap.   Page

1. THat the world lives, and is full of life.   1
2. The World hath a Spirit, Soul, and Body.   16
3. All things are made by the Spirit of the World, of the first matter.   19
4. How the Sun is called Father of the mundane Spirit and first matter.   23
5. How the Moon is the Mother, &c.   32
6. That the root of the Spirit of the World, must be sought in the Air.   37
7. How the Earth nourishes this universal Spirit.   41

The Contents.

8 The Spirit of the World is the cause of perfection in all. 44

9 The specification of the universal Spirit to bodys. 49

BOOK 2.

Chap. Pag.

That the spirit of the world assumes a Body; and how it is incorporated. 54

2 Of the conversion of the Spirit into Earth; and how its vertue remains integrally in this Earth. 67

3 Of the separation of Fire from Water, &c. 102

4 Of the Spirits ascent into heaven, and descent, &c. 160

[1]

A TREATISE
OF
The Philosophers true Salt and Secret;
And
Of the universal Soul or Spirit of the World.

BOOK I.

Chap. I.
That the World lives, and is full of life.

Purposing to comment something on the Spirit of the World, I shall first demonstrate, That the Universe is full of Life and Soul; and here besides, That Nature makes nothing Spirituous, but it also indues it with Life; and That the World consists in continual and restless alterations of forms; which cannot be without vital motion. We may also take notice, That the same Nature, like a careful as well as a fruitful Mother, embraces and nourisheth the whole World, by distributing to each member a sufficient portion of Life; so that nothing occurs in the whole Universe, which she desires not to inform; being never idle, but alwayes intent upon her action, which is Vivification.

This vast Body then, is indued with motion, yea, continually agitated therewith; and this motion cannot be wrought without some vital Spirit: for whatsoever wants Life, is immoveable. But here I mean not of violent motion from place to place; but of that, which in reference to a form, is privation; to perfection, imperfection. The vegetation of Plants, and concretion of Stones, are effected by the motion of this universal Spirit, agitating this great Mass, and the mediation of a certain radical and nutritive Spirit, whose origine or principle, like some primary procreating cause, resides in the Centre of the Earth; and thence, as from the heart, exerts all vital functions, and extends it self through the whole Body. And this root or principle is included in the bosome of the ancient *Demogorgon*, that universal Parent; whom, old Poets, those diligent Searchers of Natures Secrets, have ingeniously described, clothed in a green Cloak, obduced with rust, and covered with thick darkness; feeding all kindes of Animals, into whose belly, the vertues of the Celestial Luminaries, continually descend; penetrating the very bowels of the Earth, and impraegnanting it with all kindes of Creatures; where the elementary qualities and powers offer their services to this old Parent, as to the Producer and Distributer of all things; who continually occupies *Iliastus*, in dispensing of specifical forms; and *Archeus*, in exciting vital heat: which *Iliastus* and *Archeus*, are as it were two Instruments, whereby he informs, conserves and augments all things.

Here note, That by *Iliastus* we mean a general Steward that affords matter for all generation; and by *Archeus*, natural or radical heat, which digests this matter, and acts upon it.

This *Demogorgon* then is he, by whom, as by his Instrument, God produces all things in and under Heaven: so that he containing his *Iliastus* and *Archeus*, does with singular providence, unknown to vulgar Philosophers, and therefore masked under the supplement of occult causes, form and generate, then nourish and preserve all things; exercising the office of a good Housholder or Steward, who hath his Cellar in the bowels of the Earth, and thence draws Life and vigour for his Family. The Earth therefore, which is the Receptacle of Celestial Influences and Vertues, contains in it the Fountain of this vital Spirit, from whose Rivulets, Animals, Minerals, and Vegetables, derive Life; which communicates to them, sense, essence, and vegetation, as it findes their matter disposed for motion: and hence such things as are compounded of a more ducible Mass, and fit for such motion, become sensitive and vegetable, and able to generate things like themselves, because they are indued with Life: for Plants, and the like, whose Spirits are not cohibited in too crass and hard matter, encrease and multiply; generating things like themselves by seed and plantation; but not like Animals, but Minerals, whose Life is neither sensitive nor vegetive, but onely essential, because their composition being too hard and gross, too straightly captivates their Spirit, that they cannot produce any thing like themselves, unless they be first purged from their gross impurity, and reduced to the subtilty of their first matter: about which, *Aurelius Augurellus*, that excellent Poet and Philosopher, writes thus, *Lib.* 1. *Chrysop.*

Hæc inter variant, quæ nec primordia rerum
Extant, quæq́, frui vitali sorte negantur,
Ut media quæcunq́, sedent tellure metella
Quiq́, latent miro grati fulgore lapilli.
Nullo namq́, genus sobolemve augeri putantur

Semine; sed cunctos ævi torpere per annos;
Verum hæc ipsa etiam secreto vivere quivis
Sentiat & vitæ divino munere fungi:
Hæc & oriri eadem si contemplabitur etsi
Augeri ex sese penitus increscere cernet,
Ut mox e rariis patefactis nosse licebit.   (sunt
Quod si non sobolem educunt, non cætera ver-
In semet, causa est, quod multa spiritus illic
Materie abstrusus, vitam qui porrigit omnem,
Explicat ægre ex se vires in vivida promat
Has hominum virtus densa sub mole latentes.

But those that neither Life enjoy, nor yet
First matter are, but in the Earth do sit,
As Metals rich, and Stones that precious be,
Differ from these: for no off-spring we see
They generate, nor kinde augment, but lie
Resting themselves: yet he that shall espie
Their secret acts, and with himself compute
Their augmentation, cannot sure conclude,
That th'are quite void of life; for though they
No off-spring generate, nor turn into   (do
Themselvs, nor other things; yet life they have
Which is so chained in the closer cave
Of their dense solid matter, that they can't
Exert such actions as a brute or Plant.
Yet if mans skill do quit them of these bonds,
Their vertue's such, as will make them a-
                                          (mends.

For when these Minerals are pure, they will by their specifical form, though not generate something like themselves, yet work such an alteration and perfection in things like themselves, that they shall equalize the Philosophical Elixir, whose divine vertues, wise-men so much admire, fools so much contemn, because their blinde eyes cannot penetrate to the Centre of this Mystery.

If therefore Animals, Minerals and Vegetables, which constitute the greatest part of this visible World, be full of Life; what Reason have we to think, That the whole is more imperfect then part thereof? But let us sink further into Sublunaries: if the Celestial Bodies give Life to the Inferious, they must certainly and necessarily be en'ivened by the Universal Spirit: *for nothing can give that it hath not:* of which let us hear *Augurellus*.

Hoc etenim quicquid diffunditur undiq; Cœli,
Aeraq; & Terras, & lati marmoris æquor,
Intus agi referunt anima: qua vivere mundi
Cuncta putant ipsumq; hac mundum ducere
vitam.

All that's contain'd under Heavens Canopy,
Both fire, air, earth, & eke the boundleſs ſea,
Are mov'd, they ſay, by a moſt ample Spirit,
That th' world enlivens, & all that it inherit.

But natural motion is alwayes conjoyned with Life; how then can that produce Life and motion in another, that hath them not in it ſelf? Motion never forſakes that which hath Life; and that which either moves, or is moved alwayes, cannot want Life: The Soul of the Univerſe moving it ſelf ſpontaneouſly, is the fountain and original of all corporal motion; for the moſt ſubtile part of this Mundane Soul, ſoaring high, and inhabiting the Heavens, is continually wheeled about with the Celeſtial Bodies, which it ſelf circumduces with its proper and continual motion: and for this Reaſon, the ſuperiour Bodies are more lively and perfect then the inferiours, becauſe they are continually moved orbicularly; and that which is moved continually, muſt needs be immortal. And thus it appears, that the whole World is full of Life; and, that the Life of every species and individual, is but a participation of this Universal Life of the World, which alone may be properly called an Animal, in whose corporal Elements, the seeds of all visible and corporal thing are hidden and included: for we see many Plants grow without precedent seeds, and many Animals produced without copulation of Male with Female. The visible seed of Plants lies in their Grains; of Animals, in their Genitors: Metals also have their seed, but such as is not visible but by true Philosophers, who know how with great industry to extract it from its proper subject. And unless there be a certain procreative faculty in the Elements, wherein Generation is potentially included, many Herbs would scarce germinate on the Earth, much less on high Walls where no seed was ever sown, nor Herb planted; neither would so many kindes of Animals be generated in the Earth or Water without copulation of Sexes, as there are, which do by copulation afterwards, perpetuate their species, though themselves were not generated by the commixtion of any Parents; as we see in Snakes generated of Mud; and Flies, and other little Animals, of Excrements. Again, How do Oysters, Sea-Spunges, and other Aquatical Creatures live, which rather merit the name of Plant-animals, then of Fishes? These do not so much live by any particular Life proper to themselves, as by that Universal one, general and common to all; which is more vigorous in subtile Bodies, as more neer to it, then in grosser ones, which are more remote from it.

The World then, created wholly good, by him that is Goodness it self, is not corporal solely, but participates of spirituality and intelligence: (*for it is full of all manner of forms:*) and, as I said before, hath neither part nor member, but that's vital; and therefore wise-men have called it a Masculine and Feminine, or Hermaphroditical Animal, one part holding a certain Matrimonial Alligation with another. And hence, by a certain Translation, arises the diversity of Sexes in Plants and Animals; which, in imitation of the World, copulate together, and generate a third like themselves: for the World produces an infinity of little Worlds; for every Body in the World that is generated, is a Microcosm, having distinct parts, vertues, and qualities belonging to a little World. So that every thing hath an inclination to generate a thing like it self, by the right ordering of Action and Passion; which could not be, if all things were not full of Life: for what Generation can proceed from a dead subject, seeing it is neither probable nor possible, that that can communicate Life to another, that wants Life it self?

We see indeed sometimes, many things are generated without the congress of Male and Female, yea, without the production of either, whereinto the Universal Spirit infuses Life by means of Fomentation; as many by Artifice, who exclude Eggs and Chickins without the sitting of a Hen; and others, who by preparing certain matters, and putrefying them, produce wonderful Animals, as the Basilisk of a Cocks Egg, or of the menstruous matter of a red Hen, Scorpious of the Herb   Bees of Neats bowels, a kinde of Ducks of the Leaves of a certain Tree falling into the Sea (*a*,) with many things of the like Nature, that merit admiration rather then credit, because they are made out of the ordinary course of Nature ; certain matters in certain seasons and places attracting Life from the Universal Spirit, wherewith the World so abounds, that all its actions are vital : insomuch, that nothing dies, perishes, or ceases from action, and consequently from Life, but immediately some other living thing results out of it : and upon this account, no Body perishes, or is totally annihilated : for if it should, all the parts of the World would by little and little vanish one after another before our eyes; especially, considering how many mutations and Ages have gone before us ; insomuch, that he that perpends, might admire that there are any reliques left in Nature at this time : which a French Poet, and no little conversant in this secret Philosophy, hints at to his Friend, thus:

Vostre aspect inegal qui mea fortune change
Est comme le soliel, contraire en ses effects
Qui amollit la cire, & indurcit la fange
Et fait des corps nouveaux de ceux qu'il a defaicts.

Your aspect in my fortunes changes sways,
As Phœbus *in his effects, whose bright rays*
Waxes do mollifie, but harden Clayes,
And from corruption do sound bodies raise.

(a) Our Author seems here to embrace the vulgar Opinion of the Generation of the Northern Ducks, which the Scots call *Claikis, Claiks*, or *Claik-geese*; and the English, *Bernacles*: which many other Writers say are generated of the Nuts of a certain Maritimous Tree falling into the prolifical Sea, or of some Shells adhering to putrid pieces of Ships: which thing the learned *Lobellius* makes mention of (in *Adverf. Stirp.* pag. 456.) where first he seems to assent to, & afterwards to doubt of, and last to conclude, That *Fabius Columna* had justly refuted this Opinion. *Lobellius* in the second part of his Work, *pag.* 259. describes the figure of this Duck or Goose, as also of the

Tree and Shells. *Olaus Magnus* also mentions this kinde of Ducks, *Lib. 19. Hist. Sextent, cap. 9.*) But *Carolus Clusius* seems to have explained the generation of them more rationally, *Canctario Exoticorum*, *pag. 368*. where he sayes, That the Hollanders sayling towards *Waygatz*, saw some of these Ducks sitting upon their Eggs. *Fabius Columna* repeats his words: but *Ulysses Aldronaldus, Lib. 19. Antithog. cap. 23.* towards the end, embraceth the middle sentence, saying, He had rather erre with the multitude, then contradict so many famous writers; and therefore he sayes, These Ducks may be generated of corruption, and afterwards multiply by copulation and incubation, like Mice and other Animals. The generation of Palmer-worms from Plants, may also be well referred to this place, which, whether they be generated naturally, or artificially, feed onely upon the Herb whereof they are generated, or to which they are related: as also the Generation of Caterpillars, and then of Butterflies; which afterwards multiply their species by copulation. I saw at *Rome* in *Henricus*

Corvinus, an eximious Apothechary and Botanist, his Shop, a Butterflie, which they said was made of the corruption of Cypress-Leaves; so elegant and great, that its Wings equalized my little Finger in length, and were all over, as it were, eyed; whereof, as of the Palmer-Worm, you may read *Fabius Columna* his Observations, *Part* 2. *Stirp. minus Cog. pag.* 85.

Chap. 2.

That the World, because it lives, hath a Spirit, a Soul and a Body.

THe Body of the World lies open to our senses, but its Spirit lies hid; and in the Spirit its Soul, which cannot be united to its Body, but by the mediation of its Spirit: for the Body is gross, and the Soul subtil, far removed from all corporal qualities. For the unition then of these two, we must finde some third participating of both Natures, which must be as it were a corporeal Spirit, because the extreams cannot be conjoyned without an intervenient Ligament that hath affinity with both. The Heaven we see is high, the Earth low; the one pure, the other corrupt: How then shall we exalt this impure corruption, and conjoyn it with that active purity, without a mean? God we know is infinitely pure, and clean: Man extreamly impure, and defiled with sins. Now these could never have been conjoyned and reconciled, but by the mediation of Christ Jesus, God-Man, that true attractive Glue of both Natures. In like manner, this Spirit corporeal, or Body spiritual we speak of, is the active Glue of Body and Soul: which Soul sits in the Spirit of the World, as a spark from and of God's infinite Intelligence: for these effective elevations, renovations, mutations, variations, and multiplications of forms, must necessarily arise from intelligence, and not from matter which participates of no reason; and therefore cannot cause such formations and specifications. The World then is nourished by this Spirit, and agitated by this Soul, which is infus'd into it by mediation of this Spirit: which *Virgil*, following divine *Plato's* Doctrine, expresses elegantly, *Lib. 6. Æneid.*

Principio Cælum ac Terras composq̃, liquentes,
Lucentemq̃, Globum Lunæ, Titaniaq̃, Astra ;
Spiritus intus alit, totamq̃, infusa per Artus
Mens agitat molem & magno se corpore miscet.

The nourishment of th' earth, mountains, and
(skars
Of th' heaven, of planets, & of gliftring ftars,
We attribute to th' Spirit; but to th' Soul,
That thefe do move & ftir without controul.

To which *Augurellus* also attefts in his firft Book, faying,

Aft Animæ quoniam nil non eft corporis expers
Mundus at & mundi partes quoq; corpore conftant
Spiritus hæc inter medius fit, quem neq; corpus
Aut Animam dicas, fed eum qui folus utroq;
Participans in idem fimul hæc extrema reducat
Hic igitur Maria ac Terras, atq; Aera & Ignem
Vivereq; augeriq; atq; in fe cuncta referre
Semper Aves, femper Stirpes, Animantia femper
Gignere, perpetuamq; fequi per fecula prolem, &c.

But fince a Soul is incorporeal,
And all the parts o' th' world we meet withal
Are bodies; thefe two cannot be combin'd
Without a mean betwixt Body *and* Mind
Which is a Spirit: *wherewith the raging feas,*
Fire, air, & earth; all plants & fruitful trees
With animals, are acted; fo that they
Do generate their like, and live for aye, &c.

Chap. 3.

That all things which have Essence and Life, are made by the Spirit of the World, and of the first Matter.

ALL things are nourished by the same, by which they were produced. Now that all things breathe, live, augment and grow by this Mundane Spirit, resolve and die without it, is plain. Whatsoever therefore subsists, is made by it: and this Spirit is nothing else but a simple and subtile essence, which the Philosophers call a Quintessence, because it may be separated from gross corporeity, and the superfluities of the four Elements, and so made of wonderful activity in its operations; and it is now diffused over all the parts of the World; and through it, the Soul is dilated with all its vertues; which vertues are communicated most to such

Bodies as participate most of this Spirit: for the Soul is infused by, and transmitted from the superiour Bodies, as from the Sun, which acts most powerfully in this case: for this Spirit being calefied by the heat of the Sun, acquires abundance of Life, which multiplies and enlivens the seeds of all things, which thereby encrease and grow to a determinate magnitude, according to the species and form of each thing: upon which account *Virgil* saith,

Igneus est illis vigor & cælestis origo.

But fiery vigour, and heat celestial,
Are to these Bodies their original.

Now this Spirit is by Philosophers called *Mercurius*, because it is of many, or all forms, producing all kindes of Bodies: giving to some things a fairer, and more lasting; to others, a weaker, and more corruptible Life, according to the pre-disposition of the matter: upon which account, this fiery vigour proceeding from the Solar beams, is not alike in all subjects, but diversified as there is more or less of it in the seeds. All matters of purer pre-dispositions, have a purer and more durable Life and Spirit: for every thing delighting in that that's likest to it, it is more then Reason, that this pure Celestial vigour should penetrate and sink deeper into purer Bodies, and make them more durable and vital. For the proof of which, we need go no further then Gold, which, being purer then all other Terrestrial Bodies, participates more of that Celestial Fire, which, penetrating the bowels of the Earth, findes in Minerals the pre-disposed matter (to wit, the *Mercury* and *Sulphur*, which *Esdras* calls, *The Earth of Gold*) prepared by the action and diligence of Nature, and purged and separated from all inquinations of Terrestrial and adust Dregs: which matter, in the beginning, is onely some Sperm or Water mix'd with that Sperm, Powder or pure Sulphur; which, acted by the coagulative faculty, thickens by little and little; and in time, by long and continued action of the heat, hardens, and so comes to its perfection, which is naturally simple, tincted with the colour of Fire: for heat is the Progenitor and Parent of Tinctures. If therefore it be certain, that this heat comes from the Sun, as it must needs be indubitable, who can so much contradict Truth and Reason, as to deny the Sun to be the Author and Parent of this perfection? Let us then look higher, and seek more accurately how this perfection may be caused by this mean.

Chap. 4.

How the Sun is by Hermes *called, The Father of the Mundane* Spirit; *and of the Universal Matter.*

But some may here say, If all things proceed from one and the same matter, how can the Sun be called the Parent of this matter, when it self is procreated or produced out of this matter? For answer whereunto, we must consider, That if we speak of the primæve prejacent matter of all things, it is altogether invisible, and cannot be comprehended but by strong imagination; out of whose vital light and natural heat, this Celestial Sun was produced, with equal light and fiery vigour: which afterwards, strengthening this internal and essential heat, with natural, displayed the beams of his Fire over the whole Universe, illuminating the Stars above him, and vivifying the things below him. But because the Earth is, as it were, the common Mother of all things, the Sun acts most vigorously upon her, she being the common Receptacle of all Influences, in whose bowels the seeds of all things are absconded; which being agitated and moved by the Suns heat, come to light: and for this cause, in Winter, when the Sun is furthest absent from us, the Earth being destituted of that vigorous heat which his perpendicular Rayes brought with them, she is we see barren, and produces nothing: Whereas in the Spring, when the Sun again reviews our Climate, then she rises from her sleep or death, and receives Life and vigour. The cause of which mutation, must needs be the Universal Spirit full of Life, inhabiting the Earth principally; which, before it can generate any thing, must take up its Inn in some Body, as in the Earth, which is the Body of Bodies: and because all things are nourished and sustained by that which produces them, there must be great affinity and harmony betwixt this Spirit and the

Sun. And for this cause ancient Philosophers say, That the Sun in the Spring-time, calefies and enlivens his Parent loaden with old age, and almost killed with Winter-cold. Seeing then she is by the Sun fortified, enlivened and impregnated, *Hermes* had reason to say, That the Sun was the Father of this matter: for being otherwise barren, and without off-spring, she now conceives, generates and multiplies her spirituous matter, leading it from incorporality to corporality.

The Philosopher *Hortulanus* commenting on *Hermes* his Table, leaves and omits the radical principles of Nature; and taking his rise by the principles of Chymistry by the Sun, understands the Philosophers Gold, which he truely calls the Parent of the Philosophers Stone. For all that are conversant in this Art, learn from Experience and all good Authors, That the true matter and subject of this Stone, hath Gold and Silver in potency, and Quicksilver naturally: which Gold and Silver are much better then those men commonly see and handle, because these are alive, and can encrease ; the other are dead : and if this could not be effected, the matter would never be brought to its perfection, which this Art promises; which is indeed so efficacious, as to perfect imperfect Metals. But this same invisible Gold or Silver, which by this Magistery is exalted to so sublime a degree, cannot communicate its perfection to imperfect Metals, without the help and service of vulgar Gold and Silver. Wherefore Alchymists alway adjoyn the one or the other, and so make Gold the Father of the Elixir.

But such as would be further informed in this verity, should diligently evolve good Authors: for it is not my purpose to speak more of it : For it is enough for me, to shew, that divine *Hermes*, with one and the same finger, touches both strings, or under one and the same sentence locks a twofold meaning; which himself declares, when he asserts, That he was called *Hermes Trismegistus*, because he possessed three parts of the Worlds knowledge : for having given the Anatomy of this Universal Spirit, (which is the material Author and principle of all the three chief kindes, comprehending the whole of the world) he had attained so much of knowledge and wisdom, that nothing could lie hid from his eyes: and this principle he makes one. So that all things are produced from one, by mediation of one, and adaptation to one. This One then of which he speaks, is that general Spirit whereof I treat: and that One, by which he sayes, Miracles may be wrought, is the true Mineral matter of the Stone, whereof we spoke even now, which is produced from the first general matter, or universal Spirit, in the Earth, by Nature; which Spirit potentially containing all Celestial vertues in it self, communicates so much to this Mineral matter, as is requisite for the obtaining of its perfection.

But omitting Chymical Doctrines as much as we may in this Treatise, we say, That this general Spirit is a Stone or Elixir composed by Nature; by mediation whereof, she works all her Miracles: which is much more admirable then the Alchymists Stone, which is onely a grant. of this universal

Spirit, that it may act and perfect things like it self: for being truely Metallical, purified and compleated by Art, it purifies and digests Metals left in their impurity for want of digestion. But this Physical or natural Stone perpetually restores such things as are produced by Nature, and hourly procreates new things, as well in the kindes of Animals, as of Vegetables and Minerals; which yet it could not do, without the help and influence of the heavenly Bodies, especially of the Sun, which is the origine and principle of all faculties and generations. It hath then the Sun for its Father, and contains in it self spiritual Gold and Silver, because it is the first matter of Gold and Silver corporal. And because Air is the medium through which it receives these superiour influences, *Hermes* saith, That the winde carries it in its belly: for which cause *Raymundus Lullius* calls it *Aereal Mercury*: but the Earth, like an universal Parent, nourishes it in her fruitful womb: which appears by the production of all things proceeding from the Earth: for if this Spirit were not included therein, she would have no power nor vertue in generation and production; seeing she is properly no more then the common vessel or matrix of these many and different generations: for the general matter or *Mercury* of them, being, as Philosophers denote, invisible, and almost incorporeal, cannot be made visible and corporeal, but by some subtile artifice; which matter, if it can be extracted out of the Arms of its Nurse, and purged from all accidental superfluities, may (notwithstanding any Reason I can see to the contrary) in the things whereto it is applied, separate things corruptive and heterogeneous, and conserve and multiply things homogeneous and conformable to it.

It is without doubt, That Authors are misunderstood, when they seem to assert, that Metals onely should be usurped to the production of Metals, saying, That the Seeds of Gold are in Gold: for besides that which we have spoken of common Metals, and of those which Philosophers assume to the confection of their Magistery, we dare yet affirm, That without this general Spirit, (which is in all things the sole cause of vegetation) the faculty innate in all Metals, of becoming Gold and Silver, can never be deduced to vegetation, or from potency to act, because Nature produces not it self, but in every operation there must be some agent, and some matter subjacent to the action.

And this doubtless is that fire which *Pontanus* speaks of, which all Philosophers have concealed and kept under Lock and Key, as the sole Stearn of their actions: for want of which Fire, *Pontanus* (as himself confesses) erred two hundred times in his practise, though he had to do with the right matter.

This threefold *Mercury* then, or sum total, is the first Seed of all Metals, as also of the other two kindes or Genus"s; which is by little and little coagulated, and by the continual action of heat lying in the Myne, hardened and tincted, when it is perfectly pure: but it makes up several species, and acquires divers forms and colours, according to the variety of the place, and adjacent matter, producing Metals, Minerals and

Stones, in the bowels; Trees and Plants in the surface of the Earth; as she is animated by the Solar Rayes, without which she would be barren: for Nature at first established this for a Law, That the Sun should perpetually nourish and calefie the matter, alwayes moving its threefold faculty, Animal, Vegetable, and Mineral to its effect.

And this is the Cause why *Hermes* wrote the Sun its Father.

Chap. 5.

How the Moon *is the Mother of the* Spirit *of the world, and the universal Matter.*

LEst any might here be deceived, he must confider, That as one of us, a Microcofm, hath a Body, a Spirit, and a Soul; even so hath the Macrocofm: and seeing nothing exifts that wants thefe three, there must needs be great affinity amongft them; so that no one of them can be found without the other: and though two of them may seem sometimes to be feparated from the subject, yet they are onely hidden in the third that remains, as a subtile and profound Artift may easily experience, by the examination of Fire. What therefore is Matter, the same is Spirit; and what is Spirit, may (and that not impertinently) be called a Body. If we confider, That they are indivifible, and by Natures

Laws so generated, that they are one and the same thing: by which account it appears, that the matter is not onely Matter, or Soul, or Spirit; but refers and represents all, because one is always generated and nourished with the other: so that in the propagation and action of one, the two other are always present.

When therefore we say, That the Moon is the Mother of the Spirit and universal matter, we speak not irrationally, nor assert any absurdity: but here we must more intimously enquire, whence this Maternity proceeds. Heat and Moisture then are the two Keyes of Generation; and Heat performs the office of the Male, but Moisture of the Female. Corruption arises upon the action of Heat over Moisture, and Generation follows upon Corruption; as we may see in the small Body of an Egg, wherein, by the heat of Fomentation and Incubation, the Sperm putrefies, and afterwards the Chicken is coagulated and formed. The same is also apparent in the Generation of Man; who, by the help of the natural

Heat of the Woman acting upon the Masculine and Feminine Sperm united in her matrix, is deduced to a compleat Body, perfect in all its parts.

By Corruption here, we understand Mutation and passage of one form into another, which cannot be effected without the mediation of putrefaction, (which is the sole medium and way to Generation) which is also promoted by the help of some *Mercury* or Quick-silver, which is the special Conductor of the vegetative faculty: and the Sperms of all Bodies are aqueous, and, as it were, full of Mercurial humour: and if their innate Heat be brought from potency to act by the external heat of the Sun, then may their Generation be procured by decoction.

Hence the ancient Philosophers assert, That the Sun and Man generate: the Sun, the terrestrial Sun, which is Gold; and Man, Man. And it is manifest, That without the heat of the Sun, the heat of the Elementary Fire is dead and barren. Whence the Sun is also called the Master of Life and Generation. Heat then in all Generations comes from the Sun; but radical moisture, by the influence of the Moon; which influence, all sublunaries receive and feel, when this Planet is in its encrease or wane. And thus you have an account how and why *Hermes* called the Sun Father, and the Moon Mother to this universal matter: for the heat of the Sun, and moisture of the Moon, generate all things; because heat and moisture, in a due temperament, cause conception; and upon conception, Life and Generation. And though Fire and Water be contraries, yet one can do no good without the congress of the other: but by their diverse actions, all things conceive, and are conceived.

Ainsi dans l' univers discordante concorde
Aux Generations devient apte & s' accorde.

If Generation in the world be had,
Then what erst discord, is in concord made.

Yet I would not have my Reader suspect, That by too hasty judgement I would abstract *Hermes* his prime intenrion, from the broad way of all Alchymists into the by-paths wherein I tread; becauſe I know, That all good Philoſophers, according to his minde and will, ſay, That the Sun and Moon ſhould be in conjunction, that they may abſolve perfect Generation: for as *Arnoldus de villa nova in Flore florum*, ſays, The Philoſophers Sperm is not joyned to their Body, but by the mediation of their Moon: which Moon is not common Silver; but the true matter of their Stone, which congregates, and inſeparably retains in its belly, Body, which is the Sun; and Sperm, which is *Mercury*. And he ſpeaks of this Moon in his *novum Lumen*, where he ſays, That excepting his Maſter of whom he learned his work, he knew none that ever operated in the true matter; but all were extravagant and erroneous in the election of their matter, as if they would generate a man of a dog.

Chap. 6.

That the Root of the Spirit of the World must be sought in the Air.

Winde is nothing but Air moved and agitated, as we may learn from the respiration of Animals, which blow Winde when they breath Air. Winde then is Air, and Air is wholly vital, and the breath of Life: for without Air nothing can live or subsist: for whatsoever is deprived thereof, is suffocated and dies, yea, Plants themselves that are destituted of free Air, wither, and are in respect of others, dry and dead.

We therefore have some Reason to say, That Air is a vital Spirit, penetrating all things, communicating Life and consistence to all; binding, moving, and filling all.

By this Air then, the Universal

Spirit that lies hid and shut in all things, is generated and manifested; by this it is ingrossed, formed, and made more apt for Generation: whereof *Calid* the Philosopher treating, sayes, not without Reason, That Minerals have their Roots in the Air, their Heads and Tops in the Earth. As if he should have said, The Air causes this Spirit to enliven, augment and multiply Minerals in the Earth; though those that have some experience in preparing the Philosophers Stones, may say, That this place should be otherwise understood: for according to their Doctrine, in their Philosophical Works, there are two parts; one volatile, which is elevated in form of a vapour, and then condensed and resolved into Water; and this they call the Spirit: the other more fixed, residing in the bottom of the Vessel, which they call the Body. *Rosinus* explains this sentence by another of the same Authors: for he saith, Take the things off their souls, and exalt them on high, and reap them in the tops of their Mountains, and reduce them to their

Roots: where the Glosser sayes, These words are true and cleer, without envy and ambiguity; though he declares not what he understands by things whereof he speaks. But by Mountains (saith *Rosinus*) the Philosopher means Cucurbites; by the tops of their Mountains, Alembicks: by reaping, he means, we must receive the Water of the aforesaid things, through the Alembick into the Receptacle: by reducing them to their Roots, he means, That we should reduce the said Water to the Earth whence it arose. This is also confirmed by *Moriexus*, who saith, That the Philosophers operations consist onely in extracting Water from the Earth, and reducing it to the Earth till the Earth putrefie: for the Earth putrefies, when this Water is purified; which, being once pure, will by God's help, direct and perfect the whole Magistry.

Some have exploded Air out of the order of the Elements, thinking it as glue or lime to conjoyn divers Natures; judging it the Spirit or Instrument of the World, because it is the Chariot of the Universal Spirit: for it first receives the influences of all Celestial Bodies, and communicates them to other Elements and mix'd Bodies. In the mean while, like some divine Looking-glass, receiving and retaining the species and forms of all natural things, which it carries along with it, and insinuating it self into the pores of Animals, impresses those forms on them, whether they sleep or wake. We learn from Animals and Vegetables, That every Spirit neer the Earth, receives its vertue and vigor from the Air: for we see such things encrease and extol themselves: the Spirit which gives them Life, doth so much delight in Air, as the place where it had its origine. *Hermes* also saith, That the Air carries it in its belly. Whereunto *Aristotle* subscribes, saying, That moist things proceed from the Air, and Terrene things from the moist ones: for Air being next the Earth, humectates it on every side; and the humour thereof being condensed by innate heat, is turned into a certain kinde of Earth, which contains *Mercury* and *Sulphur* in due proportions.

Chap. 7.

How the Earth nourishes this Universal Spirit.

Though this Spirit be infused into, and dwells in superiour as well as inferiour Bodies, yet it may be best known and discerned in Bodies most evident and neer to our view; of which the Earth is neerest, and most vegetable: in it therefore is this Spirit generated, and manifested more copiously: for the Earth is a certain mark, whereto all the Influences, Rayes, and Vertues of the superiour Bodies tend. It is moreover the Fundament and Basis of the other Elements, containing in it self the seeds and seminal vertues of all things; for which cause it is rightly called the common Mother of all Animals, Vegetables and Minerals. It is therefore impregnated by the Heavens, and produces all things out of its womb ; and though this Spirit were expelled, washed away, or separated from it by what way you please, yet the Earth, thus void of Spirit, if left a while in the Air, would again be impregnated by the Celestial vertues and influences, so as to produce some Chystalline stones, and lucent sparks : and by this means, the Spirit which was taken for separated, would again regerminate in the Earth. Impregnation then made by the action of the Heavens, and of the first qualities, doth continually render her generative : for out of her womb come all things sublunary. She produces all things endued with life, preserves, nourishes, and at last resolves them into their own Nature. When she is agitated by these actions, she causes a twofold expiration ; one without her, another within her : which expirations egrede from this Terrene Spirit, when moved and calefied by the Celestial heat. The expiration elevated without or above the Earth, if it be humid, causes and produces dew and frost ; if dry, winde, thunder, and other dry Aereal impressions : but the expiration included in the Earth, if it be humid, generates liquable Metals and Minerals; if dry, stones, and the like, that are not liquable. All things vegetable proceed from, and are nourished by this Spirit, whereof the Earth is Nurse: for which cause, the ancient Poets call the Earth the common Mother and Nurse of all Creatures.

Chap. 8.

That the Spirit of the World is the cause of perfection in all.

The Universal Spirit is the general Genus, and common to every Genus: for if we cast our eyes into the inferiour or elementary World, we see it divided into three subalternals, to wit, animal, vegetable, and mineral kindes, and yet the same in all, onely operating diversly according to the diversity of its forms. And hence the infinite variety of Creatures arises; for else there would be only one species in the Universe: but if we perpend the superiour and Celestial World, we shall also finde, That the Spirit is one, equal in all, and differing in nothing but purity and subtilty: for the Celestial Spirits are procreated of its pure igneous substance, and differ from Terrene ones in corporal grosness:

and the Celestial Globes and Luminaries are made of its middle and Aereal substance: it constitutes therefore all things, because it hath in it both the faculties of superior and inferior Bodies, and because it is of an exquisite temper: for this Body is in all the beginning and end of perfection; and if it were destituted of its faculties, it could never perfect any thing: and here we understand simple and natural perfection; and although it be perfect onely, according to the intent of Nature, containing in it self the rule, line, action and power of perfection; yet it acquires vertues and faculties above the sphere of natural things, and can deduce things from potency to act. This Spirit alters and penetrates all things, though never so gross; mollifies hard things, hardens soft things, and augments, nourishes and conserves all things. This Spirit also, being in all Bodies the Author of Generation and Corruption, hath necessarily a threefold operation: for by its driness, it must enliven; and by its coldness, congeal; and by its moisture, congregate and unite: for which it hath a threefold name imposed on it, desumed from the three kindes of Earth: for they call it vitrifying, salsuginous, and Mercurial; because of *Salt*, *Glaß*, and *Mercury*, all things are made; though *Paracelsus* reckons these principles otherwise, to wit, *Salt*, *Sulphur*, and *Mercury*, adding *Glaß* as a fourth. As if he should say, All things are made of these three first principles, and reduced at last to the fourth; as though neither Nature nor Art could produce any thing beyond *Glaß*. But I shall prove my own sentence by Examples and Reason; The Bones of Animals are consolidated and hardened by vitrification, the Flesh and Nerves concreted by *Salt*, and united and congregated into one mass by the Mercurial humour. In Vegetables also, the shells of Almonds, Pine-Nuts, Wall-Nuts, and the like; as also of Oysters and Snails in Land and Sea, may be made by vitrification: and the taste demonstrates, That these Bodies are saltish; for nothing wants *Salt*, but what is insipid: yea, those things are very saltish whereof *Glaß* is made, as *Fem Kali*, and the like. Some may here object, That it is not

Glass, but *Salt*, that causes the induration of Bones, Shells, and the like, which I have mentioned. Whereunto I answer, That Experience and Reason speak the contrary: for *Salt* is resolved and melted by the least moisture of Air or Water: but the Bones and Shells before mentioned, resist liquefaction, as they are more or less hardened by this *Glass*-making faculty: for the ultimate confirmation of which my assertion, I may adduce precious Stones, *Adamant* and *Chrystal*, which are nothing but *Glass* elaborated to perfection in the Furnace of Nature. And now, that all things are condensed by *Mercury*, is so manifest, that it needs no other Testimony, but common Experience. Minerals have enough of *Salt*, *Sulphur* and *Mercury* in them: Stones and such effoded things as acquire not extention and fusion by the Hammer and Fire, have some *Salt* in them; but this is superated by the adustion of corruptive *Sulphur*, which comes upon their induration and vitrification. Metals and all ductile things are concreted and condensed by *Salt* and *Mercury*; and so much hardened

[48]

by vitrification, that they bid some resistence to the Hammer, which is indeed more or less, according to their implication with more or less impurity and adust Earth, which comes upon the coagulation of their *Mercury*. And thus we may affirm, that all things are made of the ternal number of *Glass*, *Salt* and *Mercury*, or Water; where *Glass* is the cause of hardness, *Salt* affords matter, and Water causes unition and condensation.

Chap. 9.

Of the specification of the Universal Spirit to Bodies.

THe Soul of the World, and its Action and Vertue, is represented in all things in which it is: this bindes and conjoyns the superiour things with the inferiour: for as many Idea's as the Heaven contains, so many seminal causes it obtains; whence, by the Mediation of the Spirit, it forms so many species in the matter. When therefore it falls out, that any one of these species degenerate, it may, by the Soul within it, and the mediation of the universal Spirit, be reformed, and reduced to its former state; for the Spirit is alwayes at hand, and ready for all motions. In the mean while, we must not imagine that the intellectual Idea is attracted; but rather, that the Soul is indued with such a vertue, and allured by the material forms : which cannot seem abſurd to any one ; for it prepares every one his meat and nutriment, becauſe it is tranſmutable into all things by which it is ſollicited, and willingly remains and reſides therein. *Zoroaſter* calls the Agreement and Harmony of the Forms, with the Soul of the World, Allurements. Whence it appears, That all things and kindes draw their powers and faculties from the Soul of the World, not all totally, but ſuch as reſpect the ſeed or propagation, and the like, whereby they germinate or encreaſe. An example hereof we have in Man, who, feeding onely on Man's meat, acquires not the Nature of Birds, Fiſhes, or the like, which he eats. Many other Animals alſo feed upon the ſame victuals, and yet every one attracts that which is proper to his ſpecies : ſo that it is worthy our admiration, that out of the ſame Meat, Man can attract what is proper to Man, and a Bird what is proper to a Bird. And this is not becauſe there are many and diverſe Aliments in one and the ſame diſh, but becauſe of the ſpecies nouriſhed, which attracts and changes the nutriment proper to, and convenient for it self; by mediation whereof, it generates its like, by vertue of this Soul and seed, which is in it according to its quality.

But we must not think, That in the Machine of the World, the Spirit, Soul and Body, are things separated: for these three are always united and conjoyned, as is apparent; and by this union, the whole Spirit and corporal substance become vital.

The universal Soul then, feigns and imagines divers forms, which the Spirit receiving into the bowels of the Elements, makes corporeal, and produces. Hence Animals generate onely Animals; Plants, Plants; and Minerals, Minerals: though not all alike: for Minerals, as I said before, generate not their like after the same manner as Plants, because their Spirit is cohibited by too gross matter; which Spirit, if it could be conveniently extracted, and conjoyned with Mineral matter, would generate its like, because by its exquisite penetration into imperfect Bodies, through the subtiliation of Art, and artificious Graduation of Fire; it brings with it proper and Mineral seeds onely, not animal, because repugnant to its Nature: yet I will not say, That it wants the action of other faculties; but that it doth not demonstrate them, but according to the species whereto it is accommodated: for else every thing would produce its unlike: a Tree would generate a Man; a Plant, a Bull; a Metal, an Herb: which I speak onely in respect of the diverse specifications of things.

For if we consider the most general Genus, it produces in all things its like, because, being *Mercury*, it assumes the Nature of all things wherewith it is mixed. But humane Art cannot effect that that is solely granted to Nature, which alone can procreate a species: Art may dilate and multiply it, if it begin its operation at the root of the species, as prudent Philosophers do; who, extracting the Spirit from Minerals specificated, decently purified, and reduced to perfection, render it apt to perfect imperfect things.

And an expert and industrious Artist perpending these things aright, may easily institute admirable Adaptations.

Finis Libri primi.

A TREATISE OF The Philosophers true Salt and Secret;

And

Of the universal Soul or Spirit of the World.

BOOK II.

CHAP. I.

That the Spirit of the World assumes a Body; and how it is incorporated.

WE have I hope sufficiently explicated in the former Book, that all things were not only produced; but also made corporeal by the univerſal Spirit. We now come to declare what kinde of Body this Spirit aſſumes, and how it ſelf both is, and makes all things corporeal: for that muſt needs be corporeal which makes all other things ſuch, ſeeing nothing can give what it ſelf hath not. Let us therefore ſee with what, and how it is veſted with a Body: not that we will here diſpute of the incorporation of celeſtial and ſupernatural beings, but onely of Phyſical and Sublunary Generations, and of the body of the Earth, which is the ſole veſſel and true matrix, where the prime and moſt general Incorporator of things, is it ſelf incorporated.

I ſay then, That no body can be made without a precedent Mover, that may diduce potency to act: that which exiſts not in ſave potency, may be produced into light, and according to Natures intent, reach its final term; which is, to make that a body which it would produce. Now this mover can be no other then fire or heat, which moves it ſelf firſt in the Air; for all Generations begin there, becauſe fire is the moſt active Element, and consequently most subtile and light: whence it is more prompt for motion.

Fire therefore, whose propriety is through its levity to ascend, and make unknown things visible, doth requisitely receive the beginning of its motion and action from inferiour Bodies; that is, from the centre of the Earth: where, as we said before, the old *Demogorgon*, and Progenitor of all things, inhabits; sitting there in his Throne, and midst of his Empire: that thence he may govern, command, preserve, and divide the essence of Life to all the parts of that great Spherical Body expanded about him; and that he may more easily and from equal distance, receive from every Member what he wants. The root of Fire is implanted in the fruitful bowels of this old Parent: which thence emits a vaporous breath, which *Hermes* calls the humid Nature: for vapour is the prime and next action of fire; with which it is so conjoyned, that it cannot be imagined without it.

But some may say, If vapour comes

from fire, how can it be moist, seeing fire is hot and dry? whence acquires it this contrary quality? And here we finde nothing absurd, if we do but consider, That fire cannot live and subsist without moisture, which is its aliment, support and subject; without which, fire cannot be conceived: for being by Nature active, and its action indefatigable, it must needs have something to act upon, and never be destitute of this thing.

Fire then, and co-essential moisture, are like Male and Female in all Generations, and are the first Parents of corporification of the Spirit of the World; as hereafter we shall make appear. But Fire is the first Operator, because action precedes passion, though the Patient be inseparably co-existent with the Agent; as the old Stoick *Zenon* asserts, who thought that the substance of Fire was by Air turned into Water, and there conserved, as the general Sperm, and first universal matter, whence all things were first generated. *Thales Milesius*, whom the Greeks surnamed the wise, perceiving the matter patient, said, That Water was the first matter: which *Heraclitus* attributed to the Sea: and *Moses*, more illuminated then both, saith, That *before the heaven and earth were created, the Spirit of the Lord moved on the face of the waters:* calling Fire, because of its noble and pure essence, The Spirit of God. In saying therefore, That Fire is the first principle of beings, I transgress not the limits of Reason and Verity: for it is without doubt, the first Operator, and the last Destroyer and Changer of forms, whose cause it is, even till it hath brought them to their period and first matter, beyond which there is no progress, but onely a transformation, as I shall by and by declare, by comparison with visible and familiar beings: the first active potency, which begins to operate in the production of man, is the agitation or motion of heat: which imitating the action of fire (whose Nature is chiefly separative) draws Sperm from the whole Body (wherein Man's seed is potentially included) and cocts and digests it: thereby rendring it apt for expulsion, and afterwards for Generation and Augmentation; which Generation and Augmentation is alwayes helped by Fire, which is the sole Actor: so that if it should attain the end of its exaltation, being inflamed by the Sulphur of Excrements, and impurity of Aliments, it would absume the radical Moisture, which is the Seat and Preserver of Life: which done, the fire remits not of its power and action, till by resolution and corruption, it brings the Body to Ashes; which nothing can do but fire.

But that we may make this more palpable, and finde out the first matter by the knowledge of the last, let us impose a Body on a common Fire: and we shall see, that what is inflammable in it, will be totally consumed, and redacted to a few Ashes; which Ashes also will participate of a fiery Nature; and for their last subject and matter, turn into a certain Salt, whose sole Parent and Multiplier, Fire must needs be. And though these should be further turned, yet Salt would alwayes be left, in whose internals we may finde Fire, which is delighted with its like.

And by this means, Alchymists finde, that there is something in Salt that is incombustible, or secret Elementary Fire, which hath the same actions with primitive Fire: upon which account they call it the Balsam of the Body, because it contains that that gives life; as also, that conserves and augments it; which is nothing else but a moist vapour, accompanied with moderate heat. *Johannes Fontanus* in his Philosophical Narrations, (*En son Romant Philosophiq.*) shews, That he was not ignorant of this Secret, where he brings in Nature speaking thus:

Aucuns disent que feu n'engendre
De son naturel fors que cendre :
Mais leur reverence sauvee
Nature est dans le feu autee :
Et si prover, je le vou loye
Le sel a Tesmoing je prendoye.

Some dare t' affirm, that fire can generate
Nothing but ashes; but those seem to relate
The truth thereof, who say, that in fires brest
All Natures operations are imprest.
And if hereof a proof you do require,
Salt gives you proof enough what is in fire.

And that it participates of moisture, is plain enough from its easie resolution; as also its fulness of heat is demonstrable, from its ready congelation: from which we may observe, That Fire acts, and is united with Fire; as in liquefaction Air acts, and is joyned with Air: for how in one subject could the dry imbibe the moist, if there were no innate heat, seeing driness, as it proceeds from heat, doth naturally imbibe moisture?

And hence we may easily understand, That the *Demogorgon* or central Fire, cannot be destitute of moisture on which it may act, and thence elevate a vapour mixed of two qualities, which I call, The Spirit of the World; but many Philosophers *Mercury* of *Merries*, because all other proceed from this naturally: but this elevated vapour is not yet a Body, but a mean betwixt a Body and a Spirit, participating of both Natures; which, whilst it remains in that state, can generate nothing. It is therefore necessary, That it either assume or form a Body, which it thus doth: This subtile vapour, proceeding from dry and moist principles, when it is elevated, penetrates the spungyness of the Earth, wherein it is gradually turned into Mercurial Water by the occurse of the ambient Air, and of the Earth it self; whose surface is far distant from its Centre, where the Fire resides, whence this heat arises. After the like manner, as we see in an Alembick, where the vapour or Spirit to be distilled, runs out. But this vapour and its water, partaking of two principles, heat and moisture, it is ingrossed; and by moderate and continued coction, condensed. The principal cause and mean of which action, is innate Fire, which contains this very vapour; and by its continual action, stimulates and compels it to imbibe this moisture, and to coagulate this Water: not in all parts with a like solidity and hardness, nor yet altogether; but first with a mucilaginous and different solidity. Now that which Nature assayes to do in the information of Idea's, is, to begin their induration and solidity, which must necessarily hold on in Natures way, which is a progress from one extream to another, by intermediate disposition. And Nature thus continuing its digestion, this Mucilage stays, of whose grosser matter Metals are generated in the veins of the Earth, or cavities of Rocks; which differ not in substance, being produced by one and the same seed, but onely in accidents, which they take from the diversities of the places and matrixes where they are generated. But the more subtile part of this vapour, ascends to the surface of the Earth, where it stays by compulsion; and being in continual agitation, though it can neither regrede nor ascend higher, and finding no solid matter to carry it with it, it is compelled to continue Natures intention; and therefore serves for the Generation and Corporification of individuals. But that what I have said may be better understood, let us take some one individual, and let us see how it is produced: for this will ascertain us, that the Spirit of the World assumes a Body, and shew us how it is incorporated. An Acorn may be long enough set or sown in the Earth, and consume without germination, unless some Agent be neer it, that may deduce its occult potency naturally within it, into act. And whence can any one imagine this action to proceed, unless from the central Fire issuing out of *Demogorgon*'s brest? Which Fire attracted and fomented by the Solar Rayes, will redouble its force, vertue, and efficacy. Does not then this germination derive its original from Natures Fire; which elevating and multiplying its vapour, excites the innate heat of the Acorn, which is of its own side resolved into a vapour by the mediation of Air? and this raised vapour is nourished and augmented by the first vapour, which never ceases to act on the matter of the Acorn, till it have brought it to the period of that perfection to which Nature destined it, which is, to make it an Oak: which, when it hath acquired its perfect magnitude, begins, not indeed to die, but to decline, and gradually to return to its first form, even to the Earth; where the same vapour is not idle or deficient: for of the putritude of this Tree, certain Animals, called Horse-lice, with abundance of Worms, are thereby generated; yea, when the Oak is turned into perfect Earth, it causes a new vegetation.

But if any should say, That the Acorn is augmented and multiplied to this magnitude, he erres: for it is manifest, That in this germination, the Acorn remains whole without diminution; yea, separates it self from the Tree when it is germinated. The Oak then grows not from the augmentation and multiplication of the Acorn; neither is it from addition and abstraction from the adjacent Earth: for then so much of the Earth would be exhausted, as the Trees magnitude is: it must therefore necessarily proceed from some other way & matter, seeing by these wayes it cannot. This Spirit then, or Vapour, which is ordained for no other end, is that which is incorporated, and produces this individual, from which the creation, augmentation, and preservation of all things, proceed; and not from the Mass of the Earth, which is nothing but the Excrement of the spirituous and primæve matter, as it appears by the digestion of the stomach, which rejects Excrements in the same weight and quantity almost that it assumed Meat in, when nevertheless, it hath extracted its proper nutriment from it, which is onely the Spirit included in that Mass; which, by its siccity, makes it self a Body; and by its humidity, dilates and augments it self.

Chap. 2.

Of the conversion of this Spirit into Earth; and how its vertue remains integrally in this Earth.

THe former Reasons seem sufficiently to evince, That the Spirit of the World assumes a Body: now we should declare how it is corporified; and though in this search, the labour will be great, and effect small, yet shall I endeavour to make this thing comprehensible, especially for their sakes, who in their nativities have had favourable Stars; and are thereby rendred Admirers of these rarer Effects, and Searchers of these occulter Secrets: for in that many learned and curious Men have erred in the inquisition and detection of this Body, arises from hence; That some have believed this knowledge far to exceed Man's capacity, and onely attainable by Angels or Devils. Others have thought, That it being called, The Spirit of the World, no Man should feign to himself any but an Universal Body, because an Universal Spirit must also have an Universal Body. And others have thought, That it could not be perceived, but by conversion of more perfect Bodies into their first Sperm and Spirit, by exact and industrious subtiliation: not minding nor considering, That Nature admits of no regress; and, That Bodies, by how much they are more perfect, by so much they are further distant from their principles and first corporeity.

Some are of opinion, That a Quintessence might be extracted from Bodies: perswading themselves, That the more subtile and volatile part is that Spirit they seek; so erring from the scope at which they aim, as if they would seek the East in the West: for they make Bodies spiritual, where they should make Spirits corporeal. But seeing this Spirit is manifestly converted into a Terrene Body, and without contradiction or doubt, generates all bodies; it must therefore be extracted by them, because otherwise, they forsake the direct tract of Nature: and seeing it is made a Terrene Body, it may be perfected by Fire, into something, which the Quintessentials call their Heaven. But corporification begins from the Earth: for the first work of *Mercury* is to make Earth. Why then do they begin with making of Fire? which would proceed, as if a Man should build a House, and begin at the Roof.

But such as would reduce Bodies to their first being, pretend; and have more specious Reasons, then those that would redact them to a Quintessence, save that they go in a crooked way that leads them contrary to their meanings: for, besides that Nature never goes backwards, they minde not that they should follow the way of completion, and not of destruction, or reversion to their origines and nativity: for, besides the impossibility of this course, these wayes are so long and difficult, that an ordinary term of Life cannot reach their end. Moreover, they can never attain the true and natural first Being this way, but onely a phantastical Body, far different from that wherewith Nature effects her productive operations; which is solely, the legitimate Sperm of all Bodies. And if we consider now, That all Bodies are made by Terrification, we must needs grant, That there is some prejacent subject most apt for the making of Earth. But I said, That in the beginning Fire was the first Operator in the World, which elevates a spirituous Vapour, then cocts, dries and incorporates it: for a Body cannot be made without coagulation, which necessarily follows the driness of Fire: And in what other place can this immassation, desiccation and coagulation be made, save in the Earth, whence all Bodies proceed? The prejacent matter then, must needs be there occluded; for if it be not there, then all Bodies are made of nothing; which is repugnant to Natures course, which would have every thing to have its principle; and out of nothing, nothing but nothing to proceed.

This matter or principle then is bound to the Body of the Earth, where it is nourished, ingroſſed and incorporated; and therefore, ſuch as would extract the Quinteſſence out of perfect, ſimple, or imperfect Metals, ſhould do better to open the Matrix of the Mother, and take the Sperm out there, then to kill and deſtroy her Infants or adult Off-ſpring, while they go about to reduce them to the ſtate wherein they were in time of conception. But if they ſhould open this Matrix, what would they there finde? for nothing appears to ſight or ſenſe. And many believing this way moſt profitable, have been deceived, whilſt they thought in the bellies of Mynes, to finde ſome ſeed of Gold; which miſſing of, they deſpair of their purpoſe, becauſe they never met with a middle diſpoſition betwixt hardneſs and ſoftneſs in Metals. If therefore the eye can diſcern nothing here, how then is it poſſible for them to get any thing? This indeed is a work, the other a labour. Such Searchers ſurely believe not, that the firſt matter is ſo ſubtile and looſe a Spirit or Vapour, as that it cannot be reached, ſave onely by intellect and imagination.

However, seeing it is contained in the Body of this great Parent, and dwells there, Reason evinces, That it hath something of corporeity in it: and although I have sufficiently declared to subtile intellects of what Nature it is, I shall further add, That the Pores of the Earth are full of this Vapour, which acquires a dry quality by its innate heat, accompanied with some secret moisture, by which it is condensed, and coagulated into a specifical Body: and as this moist Nature, now dried up, was first Water; so it must be reduced into Waters again by Water, which is the sole mean whereby to humectate dry things, as fire is to siccate moist things: which is a work duely observed by Nature in generating Metals: for Water flowing through the Pores of the Earth, findes there a dissoluble substance, wherewith it unites its most simple parts: to which union, all the Elements concur in due proportion.

This substance then, by its own dissolution is so conjoyned, that it of it self condenses through hardness, which is natural to it, because of its innate siccity; and by successive and long decoction, acquires a metallical induration.

But now this substance being dissoluble, of what other Nature can it participate, but of Salt? for nothing is so dissoluble as Salt; which by how much it is more burned, is by so much easier dissolved, unless it be turned into Glass.

The first matter then, is Salt; or Salt is the first Body whereby this matter becomes visible or palpable: of which Salt *Raymundus* speaks, when, in his *Testament*, he saith, We have before declared, That in the Centre of the Earth, there is a certain Virgin-Earth, and true Element, and that that is Natures work. Nature therefore is placed in the Centre of every thing: so Salt is this Virgin-Earth which hath yet produced nothing, into which the Spirit of the World is converted by vitrification, that is, by extenuation of its moisture. This is that which gives the form to all things, and without which, nothing could incur into the senses; nothing is coagulated without Salt, nothing congealed. This is that that gives hardness to Gold, as also to the Adamant and all Stones both precious and vulgar, with a Secret vitrifying Vertue. Yea, what is more, we manifestly see, that all Bodies, compounded of the four Elements, return into Salt: for if a Body putrefie, what remains but powder and ashes covering most precious Salt? and if a Body be destroyed by combustion, calcination or incineration, what rests in the last extraction but Salt? Glass-makers do give manifest testimony hereto: whence *Arnoldus de villa nova*, in his new Chymical Light, speaking of the permanent Water of Philosophers, which Water is dry, and wets the Tangents hand no more then common Quicksilver, saith, Who then can prepare this Water? Well, he may do it who can make Glass.

The same Author, speaking of the excellency of this dry Water, manifests it sufficiently, when he saith in his Chymical Treatise, which he calls his *Philosophical Breviary*, That an Operator can no more do any thing without Salt, then an Archer emit an Arrow without a cord or string. And *Fons Amantium* saith the same:

*Sans sel ne peux metre en effect
utile chose pour ton faict.*

*I can effect nothing in your rare Art,
Unleß with Salt I 'gin to operate.*

All Bodies then are composed of Salt, and, as we said before, the Principles of Composition and Resolution are equal: which concurs with the Philosophers infallible Rule, That the first matter of all things is one with the last: where they alledge Ice and Snow for example, which are by heat resolved into Water; out of which, they were by cold congealed. And if I should here suggest the Testimony of all approved Authors to this Verity, my Treatise would end in a Volume: but that I may demonstrate, That this Salt is pure and true Earth, (not such as we tread upon, which I shall hereafter prove to be nothing but the Dregs and Excrements of the other) I must recur to the first Creation, which I shall decypher by a familiar example of an Operation made in imitation of Nature, and by the same Rule and Model by which this great Universe was framed.

I said before, That *Water* or the *humid Nature*, as *Hermes* calls it, upon which *Moses* saith, That *the Spirit of the Lord moved*, was the principle of all things.

And here the Question will be, How that great and confused heap of Waters was so divided, that this ample and gross Terrestrial Mass proceeded thence; and by what medium so different things were procreated of the Earth?

I shall answer to these Questions onely what experience hath taught me: It is therefore naturally probable, that in the middle of these Waters, by way of separation, there was a certain collection of sediments or setlings; wherein I follow the Text of *Moses*, who saith, That *God separated the water from the water*: for there are two kindes of Waters, to wit, elevative and congelative Waters: the former then elevating it self in a vapour, left the other fixed in the botom; as those that coct Sea or Fountain-Salt daily experience: though perhaps it be be true, that the one is made by the attraction of the Suns Rayes, the other by the expulsion of Fire. And here note, That Fire and Heat onely are indued with a separative faculty, which they exert either by violent or natural motion. This separation then was made by one of these wayes: and to what thing could *Moses* better compare this Fire (which cannot be otherwise defined then the origine of Universal Light, of Animal heat and vital motion, which gives existence to all things, and preserves them in their being) then to the Spirit of the Lord?

Let us again consider Natures Salt in its Chaos, diffused, dissolved, and suffocated in its Water; under what form will it then appear, or with what quality will it affect our gust, but that of bitter Water? and this form and quality it would retain for ever, if it were not separated: but as soon as this elevative Water feels the action of Fire, it begins to flie from it by vaporation; and so the collection is gradually diminished, till onely a little heap of Salt be left in the bottom, which comes together as the Earth did in the first Chaos of the Universe.

And thus we see the first operation of Fire, which is the production of Driness, that is, of Earth. But as this first Earth remained still coagulated with its Excrements and Dregs by Fire; so this Salt, which is true Earth, retains its Excrements, though it seem pure, white, and full of light: for nothing is generated, nourished and augmented, but it abjects its Recrements of the formation and separation; whereof we shall elsewhere speak. Now this Salt or dry Earth, thus coagulated and setled in the Water, drinks up all its humidity, and is by the continuation of heats action, spontaneously dried: preserving all this while, its innate moisture, by which it is never deterred, and from which it hath its dissolutive vertue. After the accession of this moist and dry temperament, it is apt for production, as the action of Fire shall impel it from potency to effect: and as the Body of this great Earth hath the specifical and productive vertue of individuals; so hath that same we call Salt: not that it can produce Herbs, Metals or Animals, as the other doth; but that it conserves in its brest the original Seed of all things, as Experience by the operations of Fire, manifests, hereby giving colours, sapours, vegetations, and induration to all these kindes; and also proper Fire which the Sun hath introduced into it, whereby it enlivens and nourishes all things: which I have sometimes observed in the prosecution of a Philosophical operation, whilst I saw in this matter, without other mixtion, all colours distinctly one after another in order, and according to the internals that the Masters of this Art determine, as they should be in the matter and confection of the Philosophers Stone; together with that sudden fusion which follows upon the attainment of the highest redness, like that of wilde Poppy: but it would not produce that admirable effect in changing of Metals; but it exerted such miraculous vertues, by causing universal and natural sweats in Man's Body, that I am afraid to publish them, lest I be branded with the title of a *Scharlatane Medicaster*, though *C.V.* my Soveraign good Prince, as an irreprehensible Eye-Witness, may easily vindicate me from that injury: for when the fame of those admirable cures came to his Ears, he was pleased, like a gracious *Jupiter*, to visit the Habitation of his poor *Philemon*; induced thereto, I suppose, by the generosity of his minde, and the relation of a good Man, who was so afflicted with divers Dolours, and extenuated with the diuturnity of his grievous Diseases, that nothing but hopes on Divine Providence, or the solace of his imminent Death, could move in him resentment. The true relation, I say, of this Man, so much affected *C. V.* that he diligently took the information of many men, whom I had by this remedy restored to sanity: and if the covetousness or envy of the Man who undertook the cure of the reverend Cardinal, and my Lord's Brother *P. M.* had not prohibited the use of this Medicament; I perswade my self, by God's Grace and Benediction, it would have conduced to the sanity of many that still lie languishing for want of cures.

If therefore this Salt have all the qualities of the Earth in it, who will deny it to be Earth; or say, That it may not be called the Universal Spirit made into Earth, as *Hermes* describes it? But I aver, That this conversion cannot be effected, save by the Artifice of easie practise, but difficult perquisition: for it is without falsity, an act exceeding Man's cogitations, to render that matter visible to the eyes, and tangible to hands, which so many learned and famous Writers in all ages have thought invisible and incomprehensible; of which they have affirmed, That those that labour in this profound Theory, may be able to discourse well and plausibly of its excellency; but should never finde and know it in effect. And I profess, amongst all the curious men wherewith I have conversed familiarly this forty yeers, from which time I have had some knowledge of this matter; I can scarce finde six that know ought of it.

When then I have sufficiently declared how this Salt may be turned into Earth, which is the operation of operations;

it rests that I shew, That after conversion, its vertue remains entire. But before I enter this Discourse, I shall, as it is requisite, relate with what vertue this Salt or Spirit is of it self indued, that we may search and finde the same in it when it is converted into Earth.

I say therefore, for confirmation of my purpose, That it is without doubt, and needs no proof, that the continual motion of the Heavens, is for some end: for though Physically we may say, That the end for which a thing is moved, is, to acquire another place; yet is motion made for other causes: and the intent of this motion, is not to go from place to place onely; but so to move, as to obtain the effect of another End: for there are two Ends; the one is by Philosophers called the End for which a thing is made; as the End of *Plato*'s Generation, is for *Plato*'s Soul; and Beatitude is the End for which *Plato* studies Vertue. The other End is that to which things tend: Thus the End whereto conjunction of Male or Female tends, is Generation; but the End for which Generation is made, is a Man or an Animal: Thus the End for which *Plato* went out of *Greece* into *Egypt*, was to learn Wisdom; but the End to which he tended, was *Egypt*.

The End therefore of the Heavens motion, is not onely to acquire new places, but to influence upon inferiour Bodies: for if any one should imagine, that these influences are of no use, or that they are cast into a place where nothing can receive them, he is in an error too gross to be refuted.

This Celestial Influx is perpetual and continual, because the motion by which it descends, is orbicular; beginning in, and returning to it self. And this is the Reason why the things whereon it influences, and which it produces, are of the same Nature and quality, as without ceasing to receive the power and multiplication of these Vertues which fail not.

And seeing this Influence is not extended above the Heavens, where nothing is, it must needs be carried toward some inferiour thing, that it may act upon: for nothing is passive but what hath a Body; and what other natural Body is there in the World, but Earth? Is not this the Body of Bodies, and that solely that can subsist alone, having in it self all the qualities requisite to a Body, as Longitude, Latitude, Profundity and Superficies? Is not it the Subject and Mark that Nature hath set, whereat to aim all her Darts? where can she better accomplish her works, then on the Earth?

The Earth therefore is onely that inferiour Body that receives Celestial Influences, whose faculties and powers, penetrate, calefie, purge, separate, enliven, augment, confer, and restore.

We need not now dispute, whether the Heavens and Celestial Bodies influence upon the Earth: for experience and sense takes away this doubt. This therefore being left as sufficiently known, we shall onely declare how they make their influences vertuous. I said before, That they tend directly downwards, and not upwards; and the Earth being the Centre of this Spherical Body, they must needs fall upon her, and fasten their Points there; for the Earth is that Point of the vast World, where all the lives of these influences concentricate. And seeing the Earth is a Body so solid, that it gives solidity to all others, it is requisite that that which penetrates it be very subtile. The Heavens therefore being of most subtile matter, produce alike effects: for the operations ordinarily follow the qualities of the Body that operates. And this penetration would profit nothing, but like a Torrent running over a Field, because of its swift motion, scarce wet the surface, unless it be by somewhat stayed. But seeing it tends infallibly to the Centre, and there is stopped, because there is no lower place whereto it may descend; there it is compelled to subsist and collect it self.

Hence some have said, That the Centre of the Earth is most precious, because in it all the influences are united; which meeting there, have an infinite potency, not onely because they continually flow thither, but because they proceed from Bodies infinite, incorruptible and indeficient in vertue.

The ancient Poets, who involved their Conceits in Fables, divide the

Earth into three parts; assigning the Heaven to *Jupiter* as first-born of *Saturn*, (though some attribute the primogeniture's right to *Neptune*, and the election of the superiour Kingdom to *Jupiter*, for certain Sophistical Reasons impertinent to my purpose:) *Neptune* they make the Lord of the Sea; and that by lot: To *Pluto*, as the youngest Son, they assign the Earth for Heritage, who is yet thought the richest of the three Brethren, because in his Dominion all the Treasures of the World are contained; yea, he seems to make his two Brothers Tributaries in those things they possess as best. This Son they call, The King of Hell, and to him they give the Elysian Fields, as a delightful place, where the blessed Souls shall follow his Court after death.

Divines also assert, That Hell and the torture of Souls is in this place; adduced by this Argument: That seeing the Nature of the Stars is fiery, and all their influences concur, there must needs be incredible burnings there. Again, That place is infernal, because none is more inferiour. But that Souls should be here tormented, and that the heat of this place should be so vehement as they say, is a Solæcism in Reason, and a Contradiction to Philosophical Axioms: for besides that Souls possesse no place, according to their own confessions, and that being devested of the earthy part, and corporeal prison, they would naturally elevate themselves, and ascend by vertue of their spiritual levity, which participates more of a fiery then any other quality; they cannot, but with violence, be detained in this subterraneous place, seeing they are light; nor yet, being simple, suffer under the action of Fire, which cannot act on its like.

Why therefore they should assert, That they descend to this place to be tormented, I see no reason, unless that the burthen of their sins, wherewith they are implicated, should detrude and depress them, compelling and forcing them to the Centre of the Earth; or else that sin hath got them under its Dominion, and hath incorporated it self therewith, by some unknown kinde of composition, and so make them passible and subject, not to the simple and natural action of Fire, but to some other violent created one, destined by God to that effect: and perhaps the vertue of the Fire we speak of, is by divine Power doubled for this action; which may be proved by Holy Writ.

But I will not rashly cherish any particular Opinion, or separate my self from the Orthodox Faith; in the defence whereof, I will not onely spend my Life, but Industry, by God's assistance: yet I say in my Transient, (that I may further recede from my intended Discourse) That they conclude strangely, that say, Vehement burnings arise in this place, because the influences of the Stars concur there: whereunto I would yet yield, if they can evince, That the heat of the Stars burns and consumes like our Culinary Fire, and doth not enliven, conserve and nourish: for if it were such as they imagine, not onely the Earth, but the whole Universe had been long since burnt to ashes. Those influences indeed in old *Demogorgon*'s bowels, calefie; but not with a mortal and destructive, but with a vital heat, which there implants an uniform Vertue; which by mediation of heat, dilates it self through the whole Body of the Earth, being the first moving cause of Generations.

But we must not here conceive, That it is onely the external heat that comes from the Sun, which calefies the Earth, and causes Generation: for in Winter-time, when the Sun is furthest distant, the Earth hath abundance of heat in it, as experience shews, in Fountains, Cisterns, and profound Cells: so that in the coldest Winter-Season, Metals are cocted and indurated; yea, it is credible, that they are then most of all ingrossed, because the heat of the Centre is kept in and retained, because of the frigidity of the ambient Air. The Suns approach, and more perpendicular Aspect, is not the sole cause of vegetation in the Spring; for if so, then doubtless, as its Ascent were more sublime, Vegetations would augment more proportionably to the encrease of the heat of the Rayes: but the contrary is observed. But because every like attracts its like; and the recess of the one, causes the recess of the other; The Sun by the magnetical Vertue of its Rayes, attracts and revokes the heat of the Central Sun, detruded by the rigid cold of the ambient into the interiour parts of the Earth; which returning to the surface thereof, affects all things with a vegetative faculty.

It is not therefore the external heat of the Celestial Sun, but rather the innate heat of the Central Sun, that calefies the profundity of the Earth: for heat is twofold; the one acts by reverberation, which is external; the other by influx and penetration, which is internal, whereof I now speak; whose Nature is to enliven, augment, and conserve by the sustentation of radical moisture contained in this Fire, which I made mention of in the precedent Chapter.

But that now we may prove this Central Fire not to be so intense, as to torment and burn, let us consider how the Stars do not by their influences cause heat solely, but work other effects: for *Saturn* is cold and dry, *Jupiter* hot and moist, the *Sun* hot and dry, *Mars* hot and dry, *Venus* cold and moist, the *Moon* moist and cold, and *Mercury* all in all, participating of all qualities alike.

Whence we may gather, That all the influences are temperate, equally consisting of heat, cold, moisture and siccity; which thus meeting in moderation, cannot make the place where they meet, immoderate. The Vapour then or Spirit that comes from the Centre, participates of these four: and hence all the qualities of simples have their origine; whereof some calefie, because in them heat is predominant, others dry, because siccity superabounds; others moisten or refrigerate, according to the degree of their moisture or cold: but on the other side. Stars project many other qualities besides these, into the Centre: for they give Beings to those sapours, colours, and odours, which we taste, see, and feel in sublunary things. I say therefore, that the Stars calefie the Centre of the Earth, and that the Universal Spirit dwelling there, participates of this heat; and because it is natural to heat to separate, that separative vertue which divides the pure from the impure, the subtile from the gross, and the light from the heavy, and the sweet from the bitter, descends with these influences: which purgative separation, is the cause why all things naturally reject their Excrements, as not being of their substantial parts; which is indeed very requisite, seeing there is nothing in the World, but its Excrements, exceed its natural substance: nay, all that we see and touch, is onely the excremental part of things, that obumbrates their occult substance; as we may observe in our Aliments, whose greater Mass turns not into our substance, but goes away by the passages destined for such egress; Nature onely extracting the invisible and spiritual succe out of them, which is apt to be converted into our flesh and substance.

We may likewise affirm, That this Mass of Earth we tread upon, is nothing but the Excrement of that first substance, which was united in the Chaos; Now encompassing the Centre, and so including it in a Spherical, Equilibrial proportion, that it cannot move or fall:

for seeing it is in the lowest of places, it can tend no wayes further, unless it should ascend; which is repugnant to its Nature. But in the mean while, I do not say, That the Body of the Earth is nothing but an Excrement: for though it appear wholly excrementitious, yet there lies under its Excrements, a pure substance; which being wholly spiritual, could not become sensible without the administration of some Body; as we see in all things produced: their Seed and first Matter is invisible; but as they are carried in a corporal Mass, and excrementitious substance, we can see them; and no Body can be made without Excrements: for which cause, this substance is separated from the Body of Earth in Generations, by the influence of Celestial heat; retaining nothing of the said Earth, but as much as may be for a sustentacle for it; which from the beginning had no other use, but to serve for a Receptacle and Shop, or rather a Vessel, wherein this spiritual Matter might effect its operations; as we shall largely and plainly demonstrate in the sequel Chapter, where I shall treat of

Separation more largely.

But Separation were to little purpose, if the things separated should remain without action: Natures scope in Separation, is to enliven and abandon Death, which comes from no other cause but superabundance of Excrements suffocating the pure substance; but here I mean natural, not violent Death: but if the Seeds of things should alwayes lie buried in this excrementitious Earth, nothing would be produced or receive Life: but the Celestial Vertue by its vital influence extracts them; and these being full of Vertue, dilate and promote themselves into all and each several species, as their Nature and Composition require.

Life then proceeds from Purification, which the Stars effect by their influences, with which the augmentative and restorative faculties descend: for the Stars being in continual motion, they are continually occupied with action, and consequently with vivification, giving Life to Life; which cannot be without Conservation and Restauration: Conservation they give, by sustaining Life indeficiently ; Restauration, by restoring what the Generations of individuals consume and spend. And this we may see manifestly in the first Matter incorporated ; for being impregnated by Celestial Influences, it is of it self nourished, multiplied and augmented continually.

And hence it is called a Dragon or Serpent that preys upon it self, always regerminating ; which, where-ever it be, it takes such root, that the place shall never be quite destitute of it, though it be washed or burned; which is a certain token by which this first Matter may be known.

These then are the principal Vertues which the Spirit of the World hath, and will alwayes receive from the Celestial Influences, which produce great and admirable effects in all the Members of the Universe.

But here some may enquire, How that the first matter receiving such pure and potent Influences from Heaven, comes to be conspurcated with so many vitious qualities ; and how it retains them when it hath received them, seeing it is alwayes busied in the actions of vivification, augmentation, conservation, and restauration : for if it separate not, it will die ; and if it augment, conserve, and restore not, it will diminish, perish, and debilitate ; which it never does.

Whereunto I answer, That the Stars have a twofold influence ; one natural, the other accidental : the natural was communicated to them in their first Creation; and it is that Government of the Universe, which *Hermes* speaks of, whereby they keep it in its Being, by defending and preserving it by their influences, from destruction and annihilation ; wherewith the Spirit of the Universe is continually enriched : which applies them to, and manifests them in all things whereto it gives encrease and substance : but the accidental Influence of the Stars, is that which falls out preternaturally, by reason of their different Situations and Aspects : and this is alwayes subject to mutation, and never remains equal : and this hath only power upon the effects of Matter, not upon the Matter it self ; for whatever influence happen, though never so malign, the Centre of the Earth intermits not its actions, but absolves them as before, and produces Animals, Vegetables, and Minerals, as well as ever; and if mortifications sometimes fall out from the malignity of some Aspect, they onely touch the surface or excrementitious mass of Bodies, and not the interiour substance: and such an accident is often changed; so that the influence sometimes operates one thing, sometimes the quite contrary: which the natural and principal influx never does. Whence we may conclude, That the first Matter, as it is simple of it self, receives nothing but those Celestial Vertues, which it also retains and keeps in its Terrification.

But we must now declare, how it retains them, that we may make good that saying of *Hermes*, That its Vertue remains entire when it is converted into Earth, because all the Celestial Vertues descend and meet in the Centre of the Earth; which in their course, aim at nothing, but the information of the Matter, which is as it were the Receptacle of the supream Idea's.

The same Matter being full of Forms, is diversified not indeed actually, but potentially, into innumerable specifications: and so it is not properly a Body, but as it were a Body or Companion of a Body, whereunto it hath an Appetite, and by information moves to that end; and this motion it hath from the action of Celestial Fire, which I before called, The first Mover in the Chaos: which the old Poets, *Orpheus* and *Hesiod*, describe under the Name of *Love*: And *Ronsardus*, our French *Homer* or *Pindar*, thus delineates:

Je suis amour le grand maistre des dieux,
Je suis Celuy, qui fait mouvoir les cieux.
Je suis Celuy qui governe le monde ;
Qui le premier hors de la masse eclos,
Donnay lumiere & feudy le Cahos,
Dont fut basty cette machine ronde.

If any Rustick know not who I am,
     Love is my name :
To whom themselves, both Jove and all his train,
     Subjects proclaim.

I am the hand that moves the heavens, and th' reins
   That Stars direct;
All that the Earth and Universe contains,
   My nod expect.
When all a Chaos was, I spread my rayes
   Of light about,
And then divided that which in few dayes
   The world brought out.

Seeing then this Matter by its proper Nature and Appetite, tends to incorporation, who can rationally say, That whilst it assumes a Body, it is deprived or swerves from Nature, that by its own vertue causes corporation? and seeing when it takes a Body, it is first turned into Earth; who can deny but this Earth is indued with the same vertue? for though through commixtion and concurse, it partake of some elementary impurities, yet in its Centre it is alwayes pure: so that after its purification, Fire (which is otherwise the most active and potent Element) hath no power to destroy it, because it exceeds this Fire in perfection and subtilty. And hence it so suddenly penetrates Bodies, enlivening them, and augmenting their vertues, by restoring and preserving what is natural in them, to wit, their radical moisture, which by its fiery subtily, it purges and separates from its Excrements that suffocate it. And that I may in a word absolve all, this that most excellent Medicine, which *Siracides* saith is extracted out of the Earth, and which no wise Man will despise, It is moreover that precious Salt, whereto the Doctor of Doctors compared his Apostles, as to a most exquisite Treasure, produced by the Heavens: for he might as well have said, You are Adamants, Rubies, Pearls, Gold or Silver of the Earth; but that he knew, That all these things, though admirable, contained nothing in them comparable to this general Salt, whereunto all the rest owe the homage of their perfections. This Medicine operates like Fire, in consuming the impure, which, as heterogeneous, it disgregates from the homogeneous parts of the pure substance: and seeing the Heaven generated this Virgin in the bowels of the Earth, why should she not retain her Parents vertues; and like an Infant that naturally participates of both Paternal and Maternal Seeds, hold both Paternal and Maternal Humour? for which cause the Ancients called this progeny *Androgynos*, which is a name common to both Sexes; the Poets, an *Hermaphrodite*, as being properly neither Male nor Female, but both: and no less properly, may this Virgin be called *Uranogæa*, or Heaven made into Earth: for being Earth, she hath the Celestial Vertues tied and annexed to her indissolubly, whose admirable faculties she manifests in her operations, whereof I have given sufficient declaration in the precedent Chapters, to all such as armed with any noble spark of ingenuity, will endeavor to adventure to break through the thick woods of darkness and ignorance, or to such as *Virgil* saith, to whom Heaven has granted access or ingress into the obscure Chaos of the Earth.

Chap. 3.

Of the Separation of Fire from Water, subtile from thick; and with what industry it should be effected.

Nature, the most sagacious Agent, shews us by her proper operations, that in all things we must first consider the end for which we undertake any matter, and then finde out means to attain the end. A prudent Searcher of Natures secrets then should have perfect knowledge of the principles, progress, and qualities of matter, both internal and external, lest in his search he confound his end with his means: and forsaking the high-way which Nature hath trodden from the Foundation of the World, turn into by-paths, and phantastical unknown and unfrequented tracts.

Hermes knew the right way well; for

he was indued with the perfect knowledge of the Worlds constitution; who, desirous by Art to follow the footsteps of Nature, prudently imagined, That the Earth was the principle of all things, and the first Creature that was by Separation created in the belly of the Chaos; and thence he made such ingress into the Treasury of Natures Secrets by the Terrification of this first Matter, which I said before was nourished in the Matrix of the Earth. But as it is not enough for a Builder to have Materials wherewithal to erect an Edifice, unless he know the manner of performing and fitting his Materials to the work: so *Hermes* was not content to finde out convenient matter, but made diligent search for the manner whereby he might prosecute his work in imitation of the chief Operator in the Creation of the Universe, and so form a little World, wherein all the vertues of the greater World were included. Considering therefore, That the thing he purposed to make, must be most, and that he must begin with gross and inferiour things to obtain this perfection; he first begun to divide the contrary Natures, by separation of the superfluous and useless parts, to avoid the ruine of his work: where, according to the Adage, it may be truely said, *He took the Bird by its Feet*: and so going through the right Gate, entred into the Closet of Natures Secret, without turning: for Separation is the beginning of all things, and the first operation of the Universal Body; segregating the confused Members: by the division of the confused Mass or Chaos, the order and form of Elements first appeared, and took their places: for without Separation, Day and Night, Sun and Moon, Winter and Summer, had yet been one; Metals and Minerals, though never so different, would have been contained in one Body; and all Vegetables would have been but one Seed.

It was therefore necessary, That Nature, for the attainment of that Order and Distinction, wherewith we see the Universe condecorated, should begin with Separation. But that we may descend to particulars, let us see how this sagacious Agent begins all her operations: but for Separation, Generation would neither have beginning nor end: by separation, Aliments augment and conserve every Body. And if I should deduce the proof of this Argument thorow every species, I should without doubt confound my self in the intricacy of this *Chaos*, and never finde exiture, for the infinity of examples: I will therefore lay down this for a Foundation, *That Nature begins all her Actions from Separation.* But because the knowledge of this is not sufficient, unless withal we know what things she separates, and whence this separative faculty proceeds, we must discuss this matter more exactly, that our discourse may proceed regularly: and before I meddle with any disposition, I think it best to propose some definition of Separation, and declare of how many kindes this Separation is.

Separation then in general, is nothing but the division and distinction of things dissimilar from one another; as of the Heaven from the Earth, Sun from Moon, and the like; as also of the pure from the impure, heat from cold, dry from moist, and the contrary. And from this Definition, we may draw two sorts of Separation;

The one of things differing onely simply, and not of contraries, as of the parts of the World which were separated at first from the Chaos; or that we may descend to particulars, as of the Wood from the Bark; of Leaves from Fruits, and Roots from Boughs: and this species of Separation may be called simply a Distinction: for these parts are not really divided or separated from each other, whether we have respect to the principal or to the particular members of the World: for though Heaven and Earth seem to be separated from each other in situation, by reason of superiority and inferiority; yet they are not divided one from another, seeing there is a perpetual connexion and affinity betwixt them: as appears by many places of this Book. Whence also *Homer*, no less Philosopher then Poet, said, That the Earth was bound to Heaven by a golden Chain. But that I may keep to my former example, Leaves and Fruits, Wood and

Bark, Boughs and Roots, are not separated and divided as contraries; but onely distinguished by their proper places and garnishmeuts, whilst they have a certain affinity and mutual tie; yet so, that the one does not occupy the others place, but that they agree together, and support each other.

The second species of Separation, is the Solution or Distraction of things totally differing, contrary, or superfluous, which have no natural connexion with the substance of things; as pure and impure, hot and cold, gross and subtile, and the like: not that I would affirm, That these things cannot stand together; but, That their union and mixtion, because of their diversity, causes Destruction, or at least hinders the action of the natural Vertue of the pure substance. And this kinde of Separation may be properly called decision or division, which Nature uses in all her productions, that she may set the proper actions and vertues of every thing at liberty.

The first species is onely the distinction of parts truely dissimilar in situation and figure, but homogeneous in vertue and substance: for it is certain, That the Wood, Bark, and all the parts of a Tree, participate of one innate Vertue, which is particularly proper to the Tree, generally common to all its parts: but as to subalternate parts, there may be some difference; for some may receive more or less of the substance or vertue; but none contrary: for one and the same effect produces not things diametrically opposite, out of the same matter: a salutary Plant emits no poysonous faculty: yet that may be salutary to one Body, that's mortal to another. Thus Hellebore nourishes Quails, but kills Men. But one Plant cannot exercise contrary Powers upon one and the same subject: for Hellebore cannot both nourish and kill Quails, nor both intoxicate and nourish Man. The proper vertue of the Plant then, is in the whole Plant; and any part of the Plant may be dissimilar to another in situation and figure, but not contrary in vertue and substance: for the Leaves and Fruits have of the same substance and vertue, though one more then another,

Here some may object, That Brassica produces different effects: for according to the vulgar opinion, its decoction looses the belly, but its pulpe bindes the same.

Whereto I answer, If it be proper for the substance of this Plant to loose, it's impossible that it should also binde; for to speak plainly, the pulpe that remains after decoction, is not of the substance of the Plant, as it appears by the concoction of the stomack, which receives the substance of Brassica for Aliment, but rejects its pulpe, as an Excrement that hath no nutritive faculty in it; which faculty is wholly in the substance, and in each part of the substance: for the substance hath this propriety, That it admits of no contrary, but onely of more or less; which we must also understand of its actions and power, and not of its essence; of which we have an example: in man, no one part is more the part of a Man then another, though it may be greater, or more potent in vertue and action: the same may be seen in simples or Plants; wherein some parts are more or less hot or cold, moist or dry, as their colours and sapours manifest; yet are they not contraries; for one part of a Plant cannot kill through cold, and another of the same, kill through heat; but Experience shews us thus much, That the summities of Boughs, and that Flowers are more subtile in their Operations, then the trunks and inferiour parts of the same, because it is proper in every substance, for the more subtile and pure parts to ascend, and the impure to settle and abide next the Excrements: which Nature observes for two Reasons: first, That she may adorn the Plant; and by the variety of its digestion, make it grateful to look to. Secondly, That she may administer to Man or other Animals, what is needful for the conservation of his essence, in a greater or less degree: herein resembling a careful and provident Mother, who prepares all things necessary and convenient, each thing in its degree, as industry or possibility can render it: for she never exceeds simple perfection; and Flowers and Seeds are the most perfect parts in Herbs, which she hath elaborated: which Art beginning where Nature left, may bring to a higher perfection, by the same way that Nature kept, to wit, by Separation; as we shall hereafter further explain.

Nature therefore by this first kinde of Separation, does nothing but distinguish things, for the ornament of their subject, and use of Man, other Animals, or some part of the World, amongst which she hath sown and implanted mutual Love, and Reciprocal Affinity; so that all things do naturally, and out of a kinde of sympathy, tend to one anothers succour and service.

But the second kinde of Separation is different; for by it, Nature, or Art, in imitation of Nature, divides and segregates contraries; that is, abstracts from the substance, what is not of its essence, but rather averse to it; dwelling with it, though it be not of it; as pure with an impure thing, subtile with gross, substance with its Excrements. And this second kinde of Separation is used as the former, for two Reasons also: one is, That the pure substance be preserved from corruption and death : the other is, that it may more freely exercise its vertues and actions, as being free from its gross dregs: for when that which is impure, possesses and contains that which is substantially pure, it never ceases to oppress it, till it hath quite vanquished and suffocated it, and by that means made way for mortal corruption; which never approaches near simple and pure things, but haunts all fowl and impure ones. Now every substance is of it self simple and pure, and by consequence not subject to corruption and death; as we may observe in all superiour things, which are free from Excrements, but inferiour Bodies are otherwise; for they lie in the midst of the impure Excrements of this World, which naturally destroy and mortifie their Guests, whereas purity enlivens and conserves: corruptions and mortifications surround men, because of the multitude of Dregs and Excrements they are infected with, which make them live a short and miserable life, as loaden with Ærumnies and Diseases; so that they are here detained as a guilty Malefactor in an obscure filthy Prison, who being dubious betwixt death and hope of Life, lies there obduced with moldiness, infested with vermine, and nourished with immund and corrupt meats: for all Aliments are impure, and carry with them an Executioner, to wit, hidden poyson: whence we at length, take our own death in our own hands, and spontaneously kill our selves, seeing our Aliments have so little of enlivening and nourishing vertue in them, and that so closely enveloped with Excrements, that the digestion of the stomack, though strong, can scarce attract it: now we eating such meats, do ingest poyson with the good substance; which entring our Bodies, ceases not to be augmented and accumulated; till it obscure and extinguish the vital Light, or rather Natures legitimate Action, which is Vivification, unless some Medicine or Separation retard or suspend it.

Corruption then is induced by Excrements; and it may happen two ways: either from the Parents Seed, who being not well, and by consequence corrupt, produce corrupt Seed; degenerating more from Generation to Generation: but this is so far subject to Medicinal Corruption, as that by help thereof, the course of its activity towards Mortification may be stayed or retarded.

This Corruption may not improperly be called that Malevolent Satan, who goes about continually seeking whom he may devour: for which cause, this wanders about the Terrestrial Globe or the Excrements of the World, which have their principal seat on the Earth, which also eructate their Corruption upon other Elements: and thus Men living of them and in them, are corrupted in them and by them; and so they produce corrupt Seed, which time renders more and more corrupt: for our age being more vitious and dissolute then that of our Ancestors, hath rendered us worse then they; and the next its probable will be worse then us, and the next more dissolute still.

The second rise of Corruption is from the continual use of too too excrementitious Aliments, by which our Bodies are depraved; so that this infection will pass from the Parent to the Son, as we see in the Leprosie and other hereditary Diseases. Now these Aliments deduce their Corruption from the place where they are generated: for after the Almighty Creator had disposed the confus'd Chaos, he so ordered, that the superiour Bodies should remain pure and subtile, but the inferiour gross and impure, because it is the Nature of pure substances to ascend to the place of their origine, but of Excrements to descend to the Centre.

And hence it is, That that which is pure in Animals and Vegetables, is elevated and ascends highest, and makes them ascend & increase with it, till it be freed from the Excrementitious Mass, which subjects it to mortal Corruption, and that it may attain the place that is remotest from Excrements, and live without alteration or spot; for the same cause also more spiritual and subtile Creatures inhabit in higher places, as being purer, and finding their Aliments more convenient, and like their natural substance: but such as are coporal, inhabit lower places, where they are immerged in dregs and filth, that have their place in lower parts; whence they are corrupted by that of which they live, which is mixed with mundane Dregs: for whatever the Earth and other Elements (which are the Receptacles of impurities) can produce, is corrupt and maculated; and therefore induces corruption and defilement upon all Bodies that use it for Aliment: And thus Blood acquires an ill disposition, which afterwards beget ill humours; but in some more, in others less; according to the inquinated state of the Parents, and the abuse of the corruptive things, which cause mortality and destruction: for if the Earth and the Fruits thereof were as pure as the Heavens, all Animals would live as long as the Celestial Incolists: But Nature hath ordained this for a Law, That that partakes more of corporeity, should dwell about that that's most corporeal; and that which is more corruptible and inquinated, about that that's likest to it: but the Earth is lowest of Bodies; and therefore most gross and corruptible. Nothing therefore can proceed from it, but what is like it; unless its corruption and impurity be stayed by the Art of Separation, and all its pure substance extracted from its Body; which a true Philosopher may by industry effect.

It never was, nor yet is, my intent to offend Physitians, to whom I owe much honour; yet I, with many learned men, admire, they do not better instruct Apothecaries, that they may be more curious in preparing Medicaments; whilst themselves may observe how oft they have been frustrated in success, when they proceed after the vulgar manner: for they will cure and restore sick and weak Bodies, by offering them a great deal of Pottage; wherein there is yet so much impurities and gross dregs, that very little pure substance, wherein the curative faculty consists, remains; which is so immerged in their Poyson, that they have no power against the Disease: neither can Nature help their operation; for she is subdued in this conflict, by the impurity of the Remedy on the one side, and the cause of the Disease on another: which is all one, as if a Man should drive away Corruption by corrupted & corrupting means; which is impossible : for Rain never dries up Rivers, nor Fire extinguish a Flame ; nor yet Corruption expel Corruption. They also attempt the Restauration of their Patients debilitated strength, by Aliments of easie digestion, which they think partake of little impurity, and so are not subject to Corruption : never considering, That this way cannot much profit them ; for though they use very choise Aliments, yet are they not of much advantage, because they are indued with no action or power that can either exterminate or lessen the morbifical causes, but onely refulciate their miserable lives, almost ruined by weakness; never freeing them from death, unless, that Nature spontaneously arise and oppose the mortal Machinations of her Enemies, or else the Patient seek help from some exquisite Medicament, brought by some perite Artificers industry to supernatural purity and perfection ; which being free from Corruption, may restore the pristine vigour, and by this means, eradicate the cause of the Disease : for every true Medicament should effect two operations, to wit, purge from Corruption, and restore lost Strength; and in this the whole Basis and Art of Medicine consists, though the lesser of the two parts is onely in use, to wit, Purgation; the more excellent, to wit, Restauration, being wholly abolished, or through sloath or avarice neglected.

And that these things are thus carried, is apparent from certain Potions introsumed, which have no other effect, but onely to make the Belly laxative, or purge out not that matter that causes the Disease, but some other Excrement, that hath nothing to do with the Disease; or else through the ill preparation, dispensation, or impertinent adaptation of simples, cause superfluous evacuations, to the ruine of Nature, too weak before, but now more enervated, both by reason of vacuity, which it abhors above all things, and also by reason of violent motion caused by such Purges.

And thus they rather procure Death then Sanity; for Nature can no more endure violent motion then vacuity, and she is very impatient of these two her sworn Enemies, which attempt so evidently her destruction.

And for this cause, common Medicine seldom cures obstinate Diseases by their Compositions, as they commonly prepare them: and if amongst many, any one be cured, this happens not because of Pills, Boles and Potions; but of the strength of Nature, which is able to overcome the impure quantity mixed with these remedies, and extract something out of their pure substance, commodious for its subsidy: or else, because the poysonous quality of these excrementitious and corruptive things, is rejected and expelled by Nature; which is of that power, that it carries with it some portion of the peccant humour which is like to it, by attraction and sympathy.

And thus an extraneous Medicament, impugning the Body, moves Nature; which provoked and prepared to resist this her Enemy, rejects and violently impugnes what is contrary to her. And now, if a Medicament should always be convenient, and not contrary to Nature, it should be purged from its poyson, that is, from its excrementitious Mass.

Wherefore a Medick should first chuse such things, as are convenient for, and hold sympathy with Man's Body; and free them from their impurities, or at least chuse such as contain in them the general vertue, or innate purity; which Purification cannot be effected, till the impure and noxious be separated and destroyed, and the pure restored, which lay buried and suffocated under the Excrements.

But it being none of my Profession to make Medicaments, I will not further treat of this business; but will return into the way whence I have a little swerved. I say therefore, Seeing there is no inferiour thing, but that's infected with Poyson, immerged in Excrements, and buried in Dregs, that cause its mortification, and hinder the liberty and action of its legitimate substance; Nature is constrained by a certain necessity, to make use of Separation; which is effected by division of the pure from the impure, the subtile from the gross, and the salutary from the destructive part. But because this admirable Agent works her operations in darkness, labouring in bodies by a secret digestion, she never exceeds simple perfection; for hereto onely is her power extended: whence corporal Elements, being not able to conduce to the highest degree of their proprieties, the Bodies wherein they are included; Philosophers prudently endeavour to separate their substance from their corruptive Mass; and after this Separation in Natures way, to wit, by Digestion and Sublimation, to carry them to the highest degree of purity; getting them by Regeneration a new form, taking away their former Nature, Qualities, and Proprieties, and changing their impure Bodies into Spirits full of purity, their moisture and cold into dryness and heat: practising and effecting this, not onely in some species or simples, but in the great Body of the World also, which is our Universal Spirit: for unless the Universal Nature of all things be renewed, it is impossible to bring it to a state of incorruption. Regeneration then is the first Fruits of Separation: but as a Grain can of it self generate nothing unless it die and putrefie in the Earth, so it is impossible any thing should be renewed or regenerated, save by precedent mortification.

Mortification then is the first step to Separation, and the onely tract to that end: for as long as Bodies remain in their old Corruption and origine, Separation cannot reach them, unless putrefaction and mortification lead the way: which also our Lord Jesus divinely taught, saying, Unless that a Man, like a Grain of Corn, die, he cannot acquire life incorruptible: not as if he meant that life incorruptible should be acquired by death corporal; for then impious and wicked Men, by dying would attain Beatitude with the Good and Just: but he meant that the old Man should die, that is, mortifie and separate his old Corruption from him, which he attracted from the Seed of his first Parents.

And this Corruption is properly intemperance and excess, introduced by the eating of the forbidden Fruit, by which Death entred the World; and since which, Man ceased not to die, because afterwards the Earth and all it produced were infected by the Poyson of the fraudulent Serpent hidden in their Aliments, whose enticements led Man to transgression, and to the eating of Fruit wherein death was included.

And this corrupting Serpent is he that I call Satan, who creeping, and without ceasing going about the Earth, mixes his Poyson therewith, and with all things it produces, to wit, Animals, Vegetables and Minerals, with intent to infect the World with his Poyson, and tyrannize over Man.

Upon this intemperance and excess in diet, a privation of Vertue followed: for Vice is nothing but a neglect of what is just, and justice is a temperate desire, and continual progress to good.

This intemperance then and excess must die in us, because it generates all kinde of sin in us, and stimulates us to malice and impiety; and therefore we are injoyned to live soberly, and shun gluttony and drunkenness, the true Authors of all carnal lusts: and that we fast often, to extinguish those internal flames, which move our Sences, and burn our Blood to Corruption.

The Anatomists of Man, know, that Man is twofold; the one Celestial and immortal, the other Terrestrial and corruptible: and the first is as a Sojourner, the second as a Prisoner.

But here arises a great Question, How it comes to pass, that the heavenly Man can, whilst buried in this putrid and corrupt Carkase, conserve his essential purity? for it is manifest, That a Liquor, though never so precious and excellent, will lose its sapour and odour, if it be long contained in a stinking Vessel: and Man also, though never so sound, is subject to infection, if he dwell in a pestilent House: a heavenly Man is of himself good and sincere; but as he is joyned to the Terrestrial, to whom impurity and vices naturally adhere, he can scarcely remain free from spots.

The depravation of this essential purity, depends without doubt on the eating the forbidden Fruit; or to speak more expresly, on the intemperance of Aliments, condited with pernicious and contagious Corruption: for which cause, this intemperance & corruption must be mortified, that the old Destroyer of Man may be restrained, & that that may be repaired by newness of life, which makes us like our heavenly Father in incorruption.

And our Restorer Jesus hath taught us two ways to Regeneration: one by the Water of Baptism, and the other by the Fire of the Holy Ghost. Water washes away the spots, and Fire consumes and separates all impurity from the pure essence: and as his precious Blood (which is true Water indeed) purges away our sins, and saves us from that death, which the mortal corruption of our earthly Parent introduced; so Water dissolves and purges away those excrementitious impurities which corrupt all substances. The Fire of the Holy Spirit consumes and separates the excrementitious impurity of sins, and so the vulgar Fire blots out the impurity of Bodies; which should therefore be mortified, that it may be regenerated: and this Mortification is that Putrefaction and Digestion, which fits it to receive Separation; and this Mortification is then wrought in us, when the Sun of the Holy Spirit rises upon our Hearts, and warms our Centre, consuming by little and little the corrupt affections of old *Adam*. After the same manner, the Chymical Fire reverberating its flame about the Body that is to be purged, burns and annihilates that that is of an extraneous Nature; and that more or less, as the impurity is more or less obstinate to Separation, which is effected by Distillation.

This then is the right way which Nature observes in the Regeneration of all things; which will have no laudable effect in Medicine, unless they be gegenerated by Water and Fire. But the same things after this second Nativity are most free in their vertues and actions; whereas before, their vital functions were buried in an excrementitious Mass: and though the benign influences of Heaven had given them much vertue, yet could not they exert it: even so a Man, whilst he is detained in the impure Prison of old *Adam*, he can produce no good and laudable action.

But before I proceed further in the practical explication of these things, I must return to my begun-order, to wit, that after I have given the definition of Separation, and declared of how many kindes it is, that now I may shew what these kindes are, whence things proceed to be separated, and whence this separative vertue arises. And I have sufficiently demonstrated, That there are in Bodies two parts: the one is an Excrement, the other substance; the one accidental, and the other substantial: for substance simply considered, is wholly pure, and without Corruption; but an Excrement wholly impure, mixing it self with the substance, and contaminating and subverting its purity.

The Generation and Formation of substance is sufficiently detected in the two first Chapters of this Book: it now rests that I explain the Nature and Qualities of Excrements: where I infer, from the aforesaid Narration, that nothing but Excrements should be separated; supposing this for undeniable, That amongst passible things in this sublunary World, nothing is void and free from Excrements: for when God separated the parts of the World, he commanded, That some should descend to the lowest Region, as partaking of the grosness of the first matter, and that they should there congregate amongst themselves: and of this feculent Mass, the Earth was formed, which is nevertheless indued with some part of pure substance, though diffused through its thick crassitude: after that *Phoebus* had slain the monstrous *Python*, swelled up with poysonous humours, which was generated of the mud of the Earth; that is, when the innate siccity had imbibed the superfluous humidity, by the operation of natural heat, the Earth begun to feel the actions of this substance included in its brest: which substance is that spirituous matter which is never idle, but alwayes imployed about Generations and Vivifications; which may here be properly called the Earth, because the proper and vertuous substance of the Earth, is that alone that by proper corporation generates all Bodies according to individual Idea's; as I have described it in this Ode, out of which I shall take what is convenient for my purpose.

> *L' esprit porte sur la face*
> *De ceste indigeste masse,*
> *L' environnant tout autour,*
> *Feit separer la matiere*
> *Pesante de la legiere,*
> *Et la noire nuict du jour.*
> *Puis de l' humeur a masse*
> *Le corps plus pesant, & froid*
> *Feit la roundeur compasse*
> *Que d'un serrement estroit*
> *L' eau ou l' air contrebalance*
> *D'un poids si ferme & egal*
> *Que sans souffrir mesms mal*
> *Ne peut choir en decadence.*
> *Puis versant t' ame au dedans*
> *Et les semences du monde*
> *La feit nuisse feconde*
> *Du ciel & des feux ardens.*

When yet the world did in a Chaos lie,
The Spirit full of fiery heat did flie
About the surface of this shapeless Mass,
And so fomented it, that what erst was
One lump without all form, did now begin
By Separation, which was wrought therein

By this same Spirit, a multitude to make
Of Bodies, and likewise of forms to take
A number almost numberless: and now
Light things ascend, and heavy things below
The centre seek to pass: the day breaks out,
And all the ponderous reliques press about
The centre, where staying themselves they (make
A round terraqueous Globe, which nought (can shake
Out of its place; for 'tis on every side
Equally balanced, and cannot slide,
Seeing no over-weight presses one part
More then another: hereto the Stars impart
Their influences, which by life and seed
Make this great Nurse all things to keep and (breed.

And seeing in this Universal Separation, that which is more fiery and subtile, chuses the higher Seat; and that which is gross and massy, the lowest; it follows, that Celestial Bodies are immortal, as being most remote and separate from all immund Dregs; and that they are crooked, or excavated into rotundity, because they flew up at once: for which cause also, Nature desires a round form for eternal things, as being most perfect and indeficient: and on the contrary it follows, that Terrestrial things must be obnoxious to corruption and death, because contraries are conjoyned in their composition, to wit, Elements of contrary qualities: wherewith the impurities or dregs of the first matter were mixed; which were not created pure, as some imagine: for then all that proceeded from them would be immortal; nay, more, there could have been no Generation in the World: for without contrariety of qualities and impurities, there had been no alteration nor mutation of forms; but all would have had one face without distinction: all things would have remained equally pure and subtile; and consequently, without all ornament: yea, that I may speak freely, there had been neither matter, nor World created. It was therefore necessary, That the subtile substance should be mixed with gross dregs: for where nothing but purity is, there is no action; because there can be no action, where there is no patient, seeing what is pure hath no power over what is as pure, nor impure over as impure. But Nature being always occupied in separating the pure from the impure, for the preservation of its essence and vital encrease, hath this substance mixed with impurity, for its sole subject; which, retaining the state and Nature of its Creation, is not nourished, multiplied or augmented, save with the nutrition, multiplication and accretion of Excrements, which are not consubstantial with it, but onely Companions and Sisters of its nativity. And that these things are so, such have experienced, who by Divine Inspiration have found the way to extract this first matter, and to corporate it in imitation of Nature; which, whatever purity, cleanness, and clarity it seems to have, yet is it always loaden with a great quantity of Terrestrial Dregs, which cannot be taken away without great industry.

I think then, that I have by most valid Arguments demonstrated, That every massy Body is not nourished and preserved by the visible Earth, but by spirituous matter onely: whereby we see, that they very much abound with Excrements, and that their whole Mass is nothing else but an Excrement, wherein this spirituous Matter, that's apt for corporation, lurks invisibly: for though we eat and drink, yet all that ingredes our stomacks, goes out alomst in the same quantity that we assumed it in. We do not therefore extract our vital Oyl out of its Mass, but that pure essence and substance that lies within it.

Briefly, That excrementitious part is nothing else but the house of this pure substance, and the vehicle that carries this nutritive Spirit to the place of distribution, there to compleat its digestion and requisite Separation. Are not Trees and Plants incorporated with this mass of Excrements? and is not this Mass, the Conduit of their enlivening and vegetative Spirit, and as it were a fulciment that makes them grow? I do not say, That all that is in a Tree, or other individual, is wholly an Excrement; for in every thing, there inhabits some particle of this substance, which I cannot justly call a Body, but onely a thing apt for corporation, which Nature cannot work of it self: for though what we see and touch is progenerated of matter apt for corporation; yet this Body is not substantial, and nothing but Excrement is seen. So that Nature suffers nothing of that to appear, that is the vital essence, substance or matter of the thing; but Art, which by industry superates the simple power of Nature, can do it well enough: for an ingenious Naturalist considers this, That though the spiritous matter and substance of things be nowhere pure in natural Creation, yet being mixed with dregs, it may be separated by the digestion of the ventricle, which rejects Excrements, and retains the substance: not as if this Separation were subject to sight, but to the understanding by the appearance of the effect, whereby we finde the dregs and Excrements rejected as useless for the conservation of the essence of a Body.

Moreover, Augmentation, Restauration, and Vivification, coming to Bodies by this substance, ascertain us of this thing, though Nature whereby it acts, hides the operation: the substance then being separable, it must be innately pure and homogeneous, or similar in all its parts, but this purity cannot be manifested by Nature for her operations; for the intentional perfection of things is onely simple.

But an Artist, observing that Heat is the onely Medium and Instrument which Nature uses in the attainment of this perfection, and Fire the sole Purger and Separator, tending to perfect purification: and further minding, that in the centre of every Body, some pure substance is contained, which can be naturally separated, if not exactly, yet according to the extent of its faculties, he proposes to himself the same way, and the same Instrument that Nature uses, to wit, Fire; which he so administers, as to destroy, burn or separate the Excrements, without the destruction of the pure substance residing in the centre; which he continues, till it be perfectly pure, and the Fire have no further destructive power over it, but rather acts for its conservation, exaltation, and the introduction of a Tincture and Quality like to it, and so at length convert this whole pure substance into its own proper Nature.

An Artist then, minding this substance to reside in all things, and that all things may be burned; so that after combustion, Ashes will remain, which the Fire cannot devour, he prudently concludes, That in those Ashes there remains something not subject to the vigour of Fire and Flame. And further incumbent on his work, he findes a kinde of Salt, not produced by Fire, but rather the Victor over Fire, like the pure Gold of every burnt Body. This Salt is the last matter resting in the Anatomy of Bodies, and not Ashes; whence this Salt was ultimately extracted, and out of which, nothing can be further extracted: for if by humidity it be turned into Water, by heat it will be again congealed into Salt. Whence we may conclude, That this Water was *Mercury*, whence all Bodies were first created; and that this Water, hidden in the ashes, hinders their total consumption by combustion, as the universal *Mercury* in the Bowels of the Earth, before the production of Bodies.

Wherefore the learned *Ronillascus* in his Writings, calls this *The Water of Mercurial Fire*, because Fire generates, nourishes it; yea, augments its goodness so much more, by how much it remains longer in it: for the ultimate operation of Fire is to make Salt; and Salt is nothing but dry Water, which acquires and conserves its siccity by Fire, and thence it is of like Nature to it.

And here I would not have any wonder, That in the beginning of this Book I said, That Fire is not without Moisture; for seeing it is nourished therewith, it must also participate thereof; because all things take their nourishment from that whereof they are made. So that Fire and Moisture are as it were two Correlates; the one whereof cannot exist without the other, no not in imagination: and the Elements have without doubt, such affinity and connexion one with another, and so participate one of another, that any of them may be found in his associate: for the *Earth* contains Air, Water and Fire; *Water*, Fire, Air and Earth; *Air*, Earth, Water and Fire; and *Fire*, Water, Air and Earth: without which participations, there could be no mutual conversion, and no sympathy or convenience.

From the Premises we may collect, That nothing is destitute of Excrements, and that Excrement and Substance are the two constitutive parts of a Body; and that nothing but Excrement, as accidental, and not affine to the essence or substance, should be separated. We may likewise gather, That Fire onely procures and facilitates this operation.

But it is time that we proceed and shew how this may be done: for it is not enough to say, That Separation is the beginning of the works of Nature and Art, not to know what things are separable, unless the practise & manner of this operation be also rendered intelligible. I said before, That there were two kinds of Separation: one for Distinction and Ornament, of which I will say no more, because it appertains onely to Nature, not to Art: the other is made by Decision or Division, which I shall now explicate. I said before,

That all things consist of two parts; Excrement and Substance: Substance is of it self simple and indivisible, whether we take it general for the first Universal matter, or in particular for any species, according to the impression of the Celestial Idea, which is infinite. This Substance or first Matter in every species of compounded Bodies, is one in essence, vertue and quality: and in one and the same subject, one part cannot be of one species, and another of another: but in Excrements it is otherwise; for the better understanding whereof, I will lay the following Rule, That there are but two means or mediums for the compleating of all Separations, to wit, *Fire* and *Water:* and there are but two things separable in Bodies; the one whereof is separated by Fire, the other by Water. In the first place it is indubitable, That the Nature of Fire is such, as it destroyes and consumes all that is combustible; and of Water such, as it washes and cleanses all substance from its impurity that commaculates it: Fire devours all that's volatile, and of an Airy quality, because it is its proper nutriment: but Water divides all that's Terrestrial and gross. Betwixt these two extremes then, there must be some intermediate disposition, which should be kept and conserved, containing nothing that's combustible or impure in it, which may be subjected to these two Warriers. And it is manifest, That Adustion and Dregs are the two Corrupters and Destroyers of all things; which Divine *Hippocrates* was not ignorant of, when he said, That all Diseases proceed either from the Air or Aliments; insinuating, that the excess of excrementitious Aliments are apt to receive corruption, and Excrements by Fire exceeding natural heat, are the inflaming and corrupting causes of each Disease: for the Excrements of Aliments, fill Bodies with Terrestrial impurities, and inflammable Air generates sulphureous and adust matter, which easily conceiving hot ardours, consumes and dissipates what is radical and vital with it self; carrying it away, because the greater part of it is volatile, the less adustive. Dregs then, and Adustion, are the two Authors of corruption, and those that hinder the vigour of substantial actions in all things.

And if we require any Arguments to prove this, the stink of Digestion and Excrements demonstrates it; for that same sented in things that are burned, speaks little of good to be there. The same is also evident in the fumes of Excrements which stinking Bodies emit; for those suppose corruption; and besides the corruption they generate, they also cause these two inconveniences; the one is the hindrance of Penetration, the other of Fixation; which two actions are absolutely necessary for the conservation of life: for that which nourishes and supports life, must needs be subtile, that it may by its subtile parts penetrate Bodies, and like a secret Balsam, roborate and augment the light of life in the centre of the Body: for if it were gross, it would rather obstruct, suffocate and extinguish, then penetrate and pass through such narrow ways: and because on the other side it conserves life in its state, it should in reason be a stable, and no fluxible thing; for if it were volatile, we might every moment expect death, lest it should be introduced by the corruption generated by feculent adustion, which continually layes wait for our life. Grosness then, hinders Ingress and Adustion, Stability.

Hence we may take a good admonition for Medicine, to wit, that every true Medicament introsumed for the restoring of strength, and profligating the cause of instant death, should have two properties, to wit, to pierce to the centre of sanity, and preserve the centre, by dilating and deducing it through the whole Body; which the Ancients, with good success, revoked to practise; and not many yeers ago, the too much hated *Paracelsus*, who explained their operations to posterity, which were so long kept secret; and let him that will, affirm the contrary, I doubt not to aver, That nothing without the operation of Fire, can be brought to purity and stability; which two parts must be chiefly observed in making Medicaments: to which Assertion this valid Reason draws me, That no body in its first original and form can be truely Medicinal, as immerged in the corruption of its excrementitious Dregs, because it cannot attain the place & seat of sanity, nor yet preserve sanity, because it wants subtilty to penetrate, and fixedness requisite to the Restauration of what is corrupt and lost, and the conservation of what is restored, which cannot possibly be attained by the common way of preparation, neither in substance, nor in infusion: the impossibilty hereof in substance appears, in that such do not purge without violence, which causes rather perilous weakness, then salutary Restauration: and as to infusions, nothing can thereby be extracted out of simples, but a little introsity, which is innate in all Bodies, with a little of their excrementitious Dregs. Hence infusion attracts not the internal vertue, but onely the external sapour of things, which in their centre is quite different from that in their superficial matter: for commonly we see, all infusions are bitter; which we are forced to obdulcorate with Sugar or Honey, because most Apothecaries are not so industrious, as to extract the natural sweetness wherein the Nature of the things consists, from the things they infuse; for all bitterness proceeds from Salt, which hath sweetness at the bottom, which cannot be extracted by simple infusions, but by Fire onely, and ingenious Artifice; this sweetness being the very perfection of the Medicine.

Hence *Arnoldus de villa nova* saith, That if you know how to dulcorate what is bitter, you have the whole mystery. Which *Brucestus* was not ignorant of, as appears by his Dialogue, entituled, *Demogorgon*.

But this occult sweetness can never be manifested, unless it be quite freed from Terrestrial Excrements, and volatile and Aereal Adustion: for the earthiness thereof, gives it its extraneous sapour arising from the Excrements of Salt, from whose variety, according to the species and place wherein they are generated, the diversity of sapours doth arise: for all sapour comes from Salt; and by how much there is more Salt, by so much is the sapour more intense. And on the other side, that which is volatile and Aery, begets ill and non-natural odours, which send out a stinch, which we perceive in their burning, because of the combustion and inflammation of the unctuous Sulphur: and that volatile matter is an Excrement, is apparent from the stinking fumes of Bodies that are burning, whence smoak proceeds; that adhering to the Chimney or other part, which afterward retain the odour of the burnt Bodies, and the bitterness of the Excrements of Salt. And this is yet more manifest, by that blackness and obscurity which this fume impresses on all things it touches, which greatly hinders the light and splendor of Nature, which always desires purity and freedom from darkness, as appears by that intense splendor in perfect Bodies, a token of their purity, seeing others, more immerged in impurity, are more obscure; as Metals perfect and imperfect, & precious Stones, demonstrate.

But if we would avoid long peregrinations, and according to the oracles advice, terminate our curious journeys in our selves, by seeking out the causes of our Diseases, we should easily finde them derived from those infected fumes that obscure the light of our sanity; and thence defume a manifest sign of what is wrought within us: for a sound Man, because of the internal clarity of his natural difposition, hath a cleer and lively coloured afpect: but a sick Man, from the first onset of his disease, is pale and plumbeous, his native vigour perishing and decaying: and all these mutations proceed from the fumes of sulphureous and excrementitious Aduftion and Inflammation, which extend themselves thorow all the members, and by mediation of the pores, infect the whole Body.

This paleness may also proceed from Nature, feeling her self aggravated with a disease, and thereupon revoking all the pure and good Blood inwards, that muftring up their forces she may better oppofe her enemy, or sustain his assaults: and upon this occafion, the external parts are destitute of their clarity; and in this conflict the external parts remain as it were dead like Earth, & tend to obscurity, because Earth, wherein this battle is begun, is naturally black; as on the contrary, Fire is by Nature clear and white.

The Earth then on its part, being spisse and obscure, begets blackness and sulphureous, fuliginous, and smoaky Adustion, begets obscurity: wherefore both cause corruption, destruction and ruine: and besides these two, nothing in the world procures universal ruine; neither is any inferiour Body exempted from these evils, save Gold and precious Stones, which Nature hath brought to what perfection she can.

And so Death is to other Bodies a perpetual enemy, seeking how to destroy them; but Nature, as a careful Mother, how to preserve her works; who hath armed two potent and nimble Champions to help them, and withstand the Adversaries fury: one whereof is Fire, the expugner of sulphureous Adustion; the other Water, the expeller and driver of Terrestrial dregs. And as Nature is ingenious and subtile in all her operations, so is Art indued with equal subtilty and industry. And there's no other way to Separation instituted by Nature from the beginning, but these two, which Nature her self hath trodden from the nonage of the Universe, whose first seeds, void of form, and confusedly mixed in a Chaos, were dissolved in Water, and by the Spirit of God, the first Mover, moving upon them, divided and distinguished: whence the Separation of light from darkness proceeded, as also of distinct forms from confusion, of generations from barrenness, and of death from life: and if things in the first commixtion were thus confused, as impure with pure, Excrements with Substance, Earth with Heaven, and life with death; all were without action, power, essence and life; and the whole confusion useless.

An Artist therefore considering this, and observing, That nothing can exert its vertue, till the confusion of its purity and impurity be separated; he takes Water and Fire for his Helpers, and follows Natures copy, whose operations he should observe with all curiosity, especially in generating Metals, which are by so much the perfecter, by how much they are more digested in the bowels of the Earth. This sentence is then irrefragable, That *Water* and *Fire* are the sole general and principal means of Separation: but the composition of things being various, and one thing yielding with more ease then another, we must also act variously upon them; yet so, as not to vary from the plain path, that nature has delineated: For adustion and unctuous Sulphur, making a body inflammable and infective, may be extracted one way out of one body, and terrestrial feculency another way in others. *Calcination* and *Sublimation*, were invented to purge adustion; but *Distillation* and *Dissolution*, to remove terrestrial feculency; and *Descension*, for the conservation of weak and easily inflammable bodies: and all these are either effected by fire; as Calcination, Sublimation, and Descension; or by water, as *Distillation* and *Dissolution*.

But because both the Books of ancient and modern writers speak at large of these things, I shall not further labour in their explication, because I can adde nothing new, nor yet any ornament or facility to them: it may suffice if I adde any thing definitively, and known to my self in general; and that is, that *Calcination* is invented, for your overcoming of hard and rebellious things, and for continuity, and valid and firm composition, which hinders division, and admits not of easie separation: and hence four conveniencies arise; the combustion of impure and stinking Sulphur, the easie separation of superfluous and extraneous earthiness, the fixation of the internal Sulphur, and more prompt dissolution.

For it is natural to fire to consume those adustible parts, which are not of the essence of a substance, to facilitate the division and rejection of terrestrial excrements, to fix radical Sulphur, and to multiply salt in bodies, which alone doth afterwards admit dissolution by water.

I say therefore, that *Calcination* agrees onely to bodies, that because of their continuity scarcely yield; for spirits and volatile things, which flye away by fire, cannot be calcined, unless some fixing thing be added, which is repugnant to their nature.

Seeing there is no other end in *Calcination*, then to extract salts, wherein the better part and secret vertue of bodies consist; to which that corruptive adustion adheres, which in *sublimation* we suffer to evaporate and flie away, as useless; that the remaining substance may be better freed from terrestrial dregs, and easier educed by fire to purification and stability. It is true also, that *sublimation* requires some intension of fire; but that happens then chiefly, when the things to be sublimated are more profoundly mixed with the dregs of some other fixed body, that they may more pertinaciously hold their terrestrial impurities: and this kind of *sublimation* is most secure, save when we work upon subjects that have their dregs naturally more fixed.

Descension hath a two fold use, one is to extract Oyl out of vegetables without combustion, the other to cleanse fusibles bodies, before they be made volatile; and these are the three kindes of separation made by fire: now we shall speak a little of the other two, Distillation and Dissolution, made by water; and the former of these is made by inclination and straining, that the limpidity of bodies may with water come forth into the water of dissolved vegetables: for that same that's made by an Alembick, appertains to the order of sublimations, which is effected by elevation, and not by ablution. And though some account this for an indifferent and but a mean way, yet I say it should not be rejected, but much esteemed, as that principally which Nature uses in her operations, and which she establishes, as the sole medium for the separation of terrestrial impurity; by which also many things are brought to the height of their perfection, being by this kind of distillation, elevated and sublimated to their purity: Whence some Philosophers have call'd it a *secret sublimation*.

The second operation by water, to wit, *dissolution*, is effected by moist and moderate heat; as Horse-dung, Saint *Maries* Bath, the smoak of boyling water, or by *infusion* in water, or else by inhumation in moist places: and the end of all these is one, to wit, the reduction of the things calcined into water, that by this liquefaction, the terrestrial matter may by straining, settle in the bottom of the vessel: but this subtile practise, and all of the like kind, should be reiterated; for if any would by continual *calcination*, separate the more simple parts of the compound, and reduce to salt what hath the essence of salt; he thereby creates much loss, for the intemperate and continual violence of the fire, will sublimate and evaporate a great part of that which is with such labour sought, so that nothing will remain, but a small quantity of the soluble matter, with a great deal of dregs; and besides, that matter may by long abode in the fire, be made into glass: it is better therefore to flie to frequent reiterations, then violate nature by the excess of any precipitancy. I found such an inconvenience once in calcinating common Chryftal, which while I defired to purge from its excrements, and by long ignition to reduce to its true essence, I found both it and its excrements made into glass, and thence useless for my purpose: for though Chryftal be clear, lucid and transparent, yet in Calcination it first emits black, and then violaceous fumes, as also stinking and sulphureous odours, which sufficiently atteft its excrementitious earthinefs, as also white fumes following after the true Homogeneity of its substance, which remains clear and fluid in small quantity, till it become true chrystalline Salt; yea, by the duration of these last reiterations, its ingrateful sent is changed into a most delicate one, like that of violets-powder. By these reiterations of Calcinations, two more conveniences do yet arise:

One is, that the thing calcined by assuefaction to the fire, brings stability and subtilty to the Medicaments, as I said before.

The other, that a body often dissolved, acquires easie penetration, and ready ingression; as also more potent strength to change the state of the patient, from sickness, to sanity; from weakness, to vigour; from declination, to restauration and perfect health: and these are the ordinary ways of all separations, which have no other scope, then the purging of pure substances from excrements, and elevating of their terrestrial grossness to a fiery purity, and from imperfection to perfection: Which *Hermes* hints at, when he says, that earth must be separated by fire, and the subtile from the gross; which he says, must be done with great care and animadversion: For when he speaks of the preparation of the universal Spirit after its terrification, and opens the way to the preparation of all individuals, he would signifie, that there is something in this earth, which can scarce be retained and conserved; and this is a light and volatile spirit, which is kept by the temperament of fire.

But on the contrary, it will easily flie away with the separable and more copious volatile part, if in the operation, temperate fire, and a good method be not used with much patience; where the Artist should observe this rule with all diligence, that there are three distinct Sulphurs: whereof two are separable, to wit, the external, which perishes by Calcination and Dissolution; and the internal, which vanishes onely by decoction; and the third is fixed, which is properly the Sulphur of nature, and the proper subject of its substance: whereunto Philosophers give the name of Agent, Fixed-grain, or Element of fire.

As to the external Sulphur, it is that first volatile and adustible water; for it is plainly extraneous, and the first nutriment of fire: but the internal is more rooted in, and united to the substance; and therefore yields not, save to intent and continued heat: and therefore it assumes all colours before it egrede; beginning first with black, which is the prime sign of earthiness, adustion, and corruption, and the antecessor of putrefaction and corruption; and then passes through other middle colours, till at length it put on whiteness, which is the airy colour; and then ascends to a fiery colour, or redness, in which the power of art, and dominion of fire, is terminated, and beyond which there is no progress.

Which thing the Poets fabulously concealed under the unconstant form of *Proteus*, who turned himself into various monstrous forms, that he might affright those that would captivate him. This variety of colours proceeds from the internal Sulphur, the true author and producer of all tinctures and varieties, which are by Nature or Art observed in any subject. These colours may be also distinctly noted in the decoction of the first universal Subject, as I have above said, That it produced them in my operations: and first, whiteness presents it self, and then Natures Sulphur appears, which *Geber* says is white without, and red within: for redness immediately follows this whiteness without all help, save the continuation and augmentation of Fire; whence one Philosopher said, His Stone was a Gold-Ring covered with Silver.

I would briefly insert these few words about colours, not that I would presume to teach the preparations and observations, which I know necessary for the perfecting of that great and so much praised *Philosophers Elixir*; but onely that I may shew the curious Disciples of learned *Medea*, which induced by profound and ingenious inquisition, seek to enter the secrets of these mysterious Physicks, what Sulphurs must be rejected, and what must be retained in all things: I hope, I shall spend that time well, which I have subtracted from my houshold-affairs, if I restore any vigour or spark of life, to that languishing part of Natural Philosophy, which envious Men have buried in the Sepulchre of Calumny, under the odious title of abusive Transmutation and falsifying of Metals; though ignorance onely stop their eyes from discerning this true mystery, which they load with opprobrious contumelies, and injurious words; grounding their malice onely on the malicious lies of impudent dullards, who run about Cities, sell smoak, and obnubilate their false Sophistications under the coat of this fair Virgin; deceiving the eyes of the too credulous Commonalty with their pigments, and by their Syrenical and fraudulent Speeches, precipitating curious Inquirers into *Scylla* or *Charybdis*.

Chap. 4.

Of the Spirits *ascent into Heaven, and descent into the Earth.*

THe Almighty Creator of all things, foreseeing from the beginning, that infection and corruption growing with things compounded of Body and Spirit, would move continual and intestine wars, he opposed a remedy to this dissention; whereby the one may be conserved, and the other not destroyed: and seeing the Spirit and Substance were included in a Body, and the Body immerged in corruption, it was impossible that corruption should act upon and prevail over the Body, and yet the Spirit placed in both should be kept free, and incur no danger ; but rather, that with the Body it should yield to deaths Tyranny, which alwayes intends Natures destruction, and the prostitution of all individuals: which thing needs no proof, but appears sufficiently manifest, from the natural, and sometimes the immature end of Animals, Vegetables and Minerals, which we see every day by their corruption, when the Body being dead, the Spirit must undergo the same fortune; that is, the vertue that enlivened it, is annihilated: but because the prime Opificer would be admirable in all his works, of his meer goodness and love to Mankinde, who from the beginning he predestinated to be the Instrument of his Glory, and to whom he subjected whatever was admirable in the Creation; for his Commodity, I say, he gave certain expedient Remedies, whereby he might not onely purifie and perfect the things created, but also, preserve and arm himself against the assaults of mortal corruption.

Knowing then, that the two parts of Man were one created in another, to wit, the Spirit in the Body, and that the Body would be continually infected by corruption, and by sensuality, drawn and allured to intemperance, which infers the true corruption, and weakning of all the members; he foresaw, that the Spirit inhabiting like a Guest in the Body, could not be exempted from its contagious depravation: and we ordinarily see, That Men given to excess of intemperance and sensuality, accustom themselves to ill manners, and take liberty in all corruption, both of Minde and Spirit; neither regarding Love nor Fear to God, Honor or Respect to the World, nor Piety to themselves, nor Charity towards their Neighbours. So that it is impossible, if they be thus bound to inquinations in death, but their Spirits must undergo punishment, as they have participated of pleasure.

Seeing moreover, all mankinde by the fall of our first Parent, obnoxious to death, and thence every Man inevitably to incur total destruction and perdition; he mitigated, or rather redintegrated this Misery by an admirable Remedy, far exceeding our capacity: for knowing Man by his Spirit and his Body to participate of Heaven and of Earth, the Remedy also he made to partake of Heaven and Earth, which is competible solely to our onely Lord, Saviour, and Redeemer, Jesus Christ, who descended from Heaven into the Earth ; and by a mystery incomprehensible by us and common sense, was miraculously made Man without the abdication of his Godhead ; because our health could not come from Earth alone, corruption reigning there : but it was necessary, that the Water should come from above, where the Fountain of purity is ; he therefore came down, that he might dwell in us, and with us, and conclude us within the terms of justice and temperance ; regenerating us to newness of life, by the mutation of our Spirit and Body, and mortifying in us our corruptions and sins, and restoring us to the study of purity of vertue ; which could not have been effected save by him alone, the extream of both Natures ; for he is God-Man, that he might conjoyn superiour with inferiour things, which were dis-joyned by the incomparable distance of life and death, purity and corruption.

The Earth doubtlessly received this inestimable treasure far exceeding its merit, by a medium that cannot be comprehended, from which he again ascended into Heaven, by the Water of purification, and Fire of the Spirit, without accidents and corporal passions, though he deposed not his Body, but retained it incorruptible and glorious, having acquired immortality by death; who shall again descend from his Fathers right-Hand, into the Earth, after the Universal conflagration, to renew the World, and make a Separation betwixt the good destined to life, and the evil condemned to death.

See now how well the Omnipotent Father, the Father of all compassions, consulted for Mans good; to whose Body, joyned with his Soul, he gave an equal Preserver, whom he sent from Heaven, that he might be born on Earth; and whom, by the light of Nature, we ought to seek.

Seeing Man was therefore endued with Reason and Judgement, that he might acknowledge and comprehend his great gifts: but Man, created heavenly, for the indagation of this benefit, as too forgetful of his birth, lays out that noble and divine Light within him, in searching out frivolous and transitory vanities, and not in the pursuit of solid wisdom and verity.

Briefly, he had rather follow the inclination of his Terrestrial Geniture, then Divine and Celestial Wisdom, which he neglects as a thing indifferent, and casually sent to him from above. Wherefore the Root of Mankinde is as it were extinguished before it grow out, (except some few who have had better Stars, and more favourable Aspects in their Nativity) they desiring the possession of transitory goods, more then the attainment of Divine and precious Gifts, which our fecund Mother Nature hath publickly and in all places fixed, for the conservation of life, hurt, rather then helped, by their abundance immerged in mortal corruption: And it is apparent, That those that are of a higher Spirit, though they look upon the fulgour and splendor of those Mundane Riches, as no way despicable; yet they will not rest in this surface, but seek to Divine Vertue, occluded in the Centre, which hath indeed been cause of great errors both in Medicine and Philosophy, to such as destitute of true light, groape at, and in the dark pass by both.

Recalling then my mind to the clear light, by whose guidance we may attain that salutary and best remedy, which God ordained particularly for the conservation of mankinde, and for the obtaining of Celestial benediction; I shall endeavor with all humility and requisite sincerity, not as a Divine, but as a Disciple of Wisdoms followers, to adumbrate my conceptions in a rude style; which the lovers of verity may accept gratefully, if they finde them rational and pleasing.

I say then, that all understanding communicated to any man from man alone, is uncertain and confused, because man is ordinarily loaden with ignorance and slow resolutions; but that which he receives from the univerial Light, is clear and immoveable. For to know absolutely, is to understand a thing by its first causes; and there is no certitude in second causes, till we come to their original: wherefore we cannot know the Nature of a Species, unless we foreknow its Genus; neither can we know the Nature of Microcosmes (which are almost infinite) unless we first find out the Nature of the Macrocosme, that gives them being. Man likewise cannot be known, without the precedent cognition of the world, whose effigies he is; nor yet the great world, unless we know whence and how it was made: For how shall one know a man, whose principle is nothing but a small deformed mucilage? or how shall a man know him that is born, unless he know those that begot him? and here I mean not the second, but the first Parents, to wit, heaven and earth: and unless a man understand the first creation of these, how shall he know them? as an Embryo in a womb, which is nothing but the congregation of a certain humour, which is afterwards formed to the example of its parents, and so in progress, till it become perfect: so the Heaven, Earth, and all that is therein, that is, all this great world, is like an Embryo in the Chaos, whence none can have any light, unless he consider the first rudiments and progress of its distinction and formation.

Let us therefore first go to the Fountain and Spring, that we may thence trace the rivulets that flow from it, and know them, and by examples of forms judge of things formed.

I say therefore, that the prime and absolute creator (who is as a point whence all proceed, and an inexhaustible fountain whence infinite rivulets issue, hath a Nature proper and particular to himself, which is to conserve as well as produce the whole world: for it is the property of a good Author, to produce things; and when he hath procreated them, to conserve and defend them.

This first Effect, which is Creation, is a secret that we are ignorant of; we understand it not, but by the affinity or likeness it hath to generations.

But the second is open and manifest to such, at least, as are illuminated, elect, and born of the Spirit; but not to such as are sons of the flesh, lest precious pearls should be trod upon, and cast to swine.

Jesus Christ our Lord hath perfected the former and more excellent, and taught us, commanding us to imitate him in all good works, whereof he hath set us an ensample: For Nature walks always in the same tract, and never forsakes her ways. As therefore the universal Parent or Conserver, consulted by his providence the common conservation of all things, from the beginning of the world: So Nature from the beginning had her intentions, and was always occupyed in continual action about productions: For as it was necessary that the safety of the Spiritual part of man should descend from above, so is it likewise as necessary, that the safety of the Body should come from the same Fountain, because from below, where the seat and habitation of corruption is, neither life nor safety can proceed. For what end do the Heavens, the perpetual Fountain of Restauration and Perfection, by the influence of their vertues flow upon the Body of the Earth, which the benevolent Stars, with their benign Aspects, sympathizing with the afflicted society of mankinde, do daily graciously prosecute and affect, but to generate in her a durable and enlivening Spirit, which in the Womb of this fecund mother assumes a Body, and manifests and dilates its faculties through all the parts of the world, distributing them to each creature according to its exigency: and hereupon do the particular powers depend, which we must know, by their effects in Herbs, Beasts, Stones, and other things which have drawn their proprieties from this general Spirit, and do miraculously conserve us and other creatures. And as it pleased God to enrich man with the perfection of his own Son, according to the extension of the humane Nature, and yet he would not have any one contaminated with sins and iniquities, to seek remedies and health from man, but from himself the true Fountain, whence all perfection flows: so also Nature, the perfect observer of the Divine will, and imitator of his works, hath not setled the perfect vertue of curation and restauration, in Herbs and particular Creatures, but would have us seek it precisely in the Centre, whence this is communicated to Creatures, to wit, in the Earth, where this enlivening Spirit is generated: for if simples are indued with the vertue of curing, restoring, nourishing, and conserving how much more hath the abundant dispensor of it, whence all these receive it? but that the Earth is the treasury and dispensor of these vertues, daily experience gives us suffient reasons; for she must needs possess them, else she could not give them.

It is therefore admirable, that so many egregious men should spend their time in drawing their waters from simple rivulets, far removed from their pure Fountain, and running through a muddy current, and not go to the Fountains head.

I do not in the mean time contemn special Medicaments, but I would that Generals were more sought after, and particulars not neglected: for though a General Medicament might salve every Sore, yet particulars merit commendation; especially in external superficiary Diseases, when the Centre of sanity is not afflicted.

That therefore I may return to my Scope, I say again, the Earth is the Matrix, wherein the Heavens beget that Spirit, that Nourisher, Restorer, and

Conserver of Bodies, which alone gives solidity and perfect cure: and how this so potent and efficacious a Spirit should be found and taken, all wise men should direct their cogitations to so useful an inquisition, that they might observe the exemplary Steps, which Nature paces in perfecting her intentions; and hold this for a rule, though God infinitely exceeding Nature, is not bound to natural reason, no more then a Monarch to the Laws himself prescribes, which yet his subjects observe without inquiring why he prescribed them: And who hath more faithfully and successfully followed this example, then *Hermes Trismegistus?* who after the deluge, was first (as some say) that opened to men the mysteries of the perfect knowledge of God, and exactly explicated Natures secrets: for besides that Angel-like in his *Pœmandrum*, he explains Divinity, where he manifests the Doctrine of the Creation of the great and small world, its beginning, progress, and duration; he also continues his holy Philosophy with the same hand in his *Asclepium*, declaring with a spirit and voice prophetical, that the regeneration of man should be wrought by the Mediation of the Son of God, indued with humane flesh: he also industriously touches the same Scope in his *Tabula Smaragdina*, where he saith, that as all things in the world were made out of one subject, by mediation of one God, so his Magistery, which is the chief and general Medicine, may be perfected and compleated by adaptation onely; which adaptation is nothing else but a Glass, where we may see Divine Meditation ænigmatically represented, to shew that Nature necessarily follows the steps of her Master.

He also attests in other Books, That the Author of Regeneration should descend from Heaven, and become Man, and live among Men for their edification. He says also in his *Tabula*, (which he left as the last Testament and Testimony of the excellency of his thoughts) That this general Spirit, the Conserver of Bodies, shall descend from Heaven, to wit, from the Sun and Moon (which in his *Pœmandrum*, he calls, The principal Rulers in this Mundane Monarchy) to assume a Body in the Earth, which he calls the Spirits Nurse, by the Mediation of the Air, in whose Belly he shall be carried, because the Celestial Influences cannot be communicated to the Earth, unless the Air, which is like an Intercessor, carry them to it: and as the Divine Restorer and Protector of our Souls, in the assumption of humane Flesh, deposed nothing of his Divinity; so (saith he) the Universal Spirit shall retain and keep all its Vertue entire, when it is turned into Earth, that is, when it assumes a Terrestrial Body.

God also willed his own Son, and our Redeemer, in his assumed Humanity, be as it were regenerated by and with the Water of Baptism, and Fire of the Holy Spirit: not that he needed any such purgation, but because he was to converse with Men contaminated with corruption, that he might in all things yield himself an example of Renovation and Purification; giving Men a visible and ample Testimony, That, according to the Flesh, he was of the same Nature with them: not contaminated or corrupted, but obnoxious to passion, and equally mortal as they.

In like manner, our fecund Mother Nature, willed her first-born, though pure in his Centre, That he should be regenerated by Water and Fire, that is, by Separation of the Terrestrial from the fiery part, the spisse from the subtile, and the impure from the pure: which *Hermes* also aimed at, when he said, That the Earth should be separated from Fire: for what God hath conjoyned, Man should not separate; but onely the impure and gross from the subtile and pure by Fire.

And besides this Sense, which first offers it self to our Intellects, there is yet another more occult meaning: for seeing by the Separation of Earth from Fire, he means, that that's gross, from that that's subtile; he seems to hint, that we should separate the natural qualities of these two Elements, by detracting the moist frigidity mixt with heavy Terrestrial things, without which that cold cannot subsist, and by putting on hot siccity, which is of the Nature of Fire, and consequently light and spiritual.

For which cause, he adds, That it must ascend from Earth to Heaven, that is, from imperfection to perfection: for *Paracelsus* calls Fire the Firmament: and as nothing can attain perfection, unless it first depose its imperfect, gross, and mortal bark, wherein this cold quality abounds, the cause of all mortification, as heat is of life; so also, prudent Nature observes this Rule, That each subject should sustain and pass the obsure blackness of death, and expect the cleer and candid renovation of life and immortality; that is, impassible essence, whereon neither Fire nor corruption hath any power. And assuredly, the acquisition of this life by death is naturally exercised by all Creatures continually; for all Sperm or Seed of Animals is mortified in some Matrix, and of Vegetables in the Earth, before any specification can be made. But if this rule takes place in the members, by how much may we more exactly observe, and directly imitate it in the head? and if the life acquired by mortification, be of any duration, how much more shall that be perpetual, which is principal?

Jesus Christ taught these things by similitude of a Grain; which he says cannot fructifie, unless it first die; signifying the mystery of his Resurrection, which his Death should precede: for he willed Death, that he might rise to a glorious and perpetual Life: therein not onely giving example to men, but expressing the whole Idea of Nature.

That divine and learned Ermite, *Morienus Romanus*, who is often quoted by modern and natural Philosophers, avers the same of the fixed Grain, whereto Nature hath given power to perfect Metals: for he saith, Unless it putrefie and grow black, it cannot be perfected and compleated, but returns to nought.

I have taken liberty to say thus much, that I may teach the younger sort, how the Creator should be acknowledged by simple Creatures: and because the vulgarity of Men begs this knowledge from remoter things; acting as they do, who seek the perfection of Sciences, from the Scholars of the lowest Form, whereas they should require it from the best Directors and Doctors; I would excite them by these natural conceptions, that they would convert the principal endowments of their Souls, to the search of the general principle; and that in more exquisite things, such as impart life and preservation to all mortal Creatures.

Mortification then, necessarily precedes all entrance into life, and principally in this Spirit the first-born of Nature, when it assumed a Body; for else, no Man could separate it from Body, which hinders its Regeneration to Life, and Pacification of its Essence: not, as though by combustion and destruction it lost its Body in Death, nor yet by Putrefaction; but so, that in Germination, the Putrefaction of Seeds annihilates not that which is corporified in them: for which cause, in the Exaltion of *Mercury*, or the Universal Spirit, after the first degree, which is made by Separation, all that's corporeal and spiritual becomes volatile, because the sublimatory vertue therein, overcomes the fixing faculty; but the fixed part afterwards retains the volatile with it, being helped by the action of heat;

which augmenting the power of the two nobler Elements, destroyes the power of the two weaker : which *Hermes* hints at in a certain Treatise, by a plumous Bird detained with a Bird without Feathers: And *Nicholaus Flammellus*, by two Dragons, one with and another without Wings.

But that I may not longer fold my self in these Dædalean Labyrinths, see we not all Vegetables encrease, and elevate themselves upwards by vertue of this volatile Spirit? which, as I said before, would carry them higher towards the place whence it came, whereto it hath an appetite, but that it is detained by its proper Earth and corporal Mass, wherein some fixed matter resides.

But lest some, not sufficiently accustomed to Philosophical terms, should think we contradict our selves, I will here explicate my self : I say then, That I mean not by this volatile Spirit, that which I before called volatile and separable Sulphur ; for that is rather the Author of Corruption then Growth : but that most simple part of the primæve vapour, which never loses its subtily, whose Nature is to be elevated, and tend to perfection; for to sublimate, is properly and Philosophically, no more then to perfect and exalt matters from imperfection to perfection. As therefore this *Mercury* hath an elevable; so hath it a stable substance: the first is naturally innate in it; the second is in its centre or power, yet it cannot compass its effect without the help of Art. And that I may shew more plainly, in what ways Nature proceeds in her operations, I think it convenient to add a little about the causes and manner of Fixation.

Repeating therefore that indubitable Axiome I alledged in the beginning of this Book, as perpetually observed in the Worlds Constitution, to wit, That all that hath life, hath also some duration; and that nothing is produced under the cope of Heaven, which hath not some kinde of life in it: I say, that this Duration must be wrought by Conservation to Perpetuity: for Perpetuation is the scope of Nature; seeing it is the endeavour of every good Opificer, to preserve the work of his hands, till it be corrupted by the injury of time, or the light of its life extinguished by the cold ashes of death, to whose feet all things necessarily prostrate themselves, by this inevitable law, That whatsoever hath beginning, must also have an end. For if all things should remain in their first extream, that is, in their beginning, without progress to their second extream, that is, their death, all things had yet been left in their Chaos; or rather, nothing would have consisted in its being; and the principles of all subjects were useless, and destructive to themselves.

Nature therefore, to eschew these inconveniences, observes the said order and progress of things, existing in continual action and motion, that is, conservation, and perpetuation.

And now, that which extends life, or conserves it, cannot subsist against the force of destruction, without some fixation and constancy; and this conservative essence, is in some more fixed then in others; whence they are also of a longer and more durable life, and more difficultly destroyed and mortified; as a

Hart and Crow, amongst animals; an Oak, amongst Plants; and Gold, amongst Minerals: and this happens by the more equal and digested commixtion of Elements: so that death, whose property it is to divide and disjoyn, cannot so easily enter these compounds, as being firmly united and well digested; and by how much Bodies are more firm thus, by so much they are less subject to the accidents of mortal corruption.

But Nature being not able of her self to attain this perfection of union and digestion, cannot totally and finally save and preserve Bodies from destruction; but the industry of Art (though Art of it self be nothing without Nature) imitating her in these things, exceeds her in the proper course of her own ways: for observing that conservation and prolongation of life, is attainable by something tending to fixation, which must be effected by union and digestion (for nothing can be fixed but what is Homogeneous, and of one Nature) the Artist labours, that he may find out the thing that is fixable, and deduce it to perfect fixation; which he doth by the same ways, order and operation that Nature uses, to wit, by separating extraneous, and uniting Homogeneous parts, which he absolves by long and ingenious digestion of the things united.

But because it is impossible for him to separate or extract this from individual and specifical Bodies, because of their firmer union and more compact digestion, he is glad to seek it in the bowels of the Earth, whence all things proceed: for to extract it entire and absolutely vertuous from another place, were a work of no profit, and impossible; and to think how it may be made perfect, is a labour both long and dubious: whence the Poet said well,

Hic sive nullibi illud est quod querimus.

Here or nowhere is that we seek.

And they are doubtlesly deceived, who following crooked and by-paths, stick in the common signification and rind of Philosophical words, and study not to find out the lively marrow of their intentions: They should therefore sacrifice first to the infernal Queen; for there is the Fountain and Spring of all things.

Wise men begin their works from the root, and not from the branches; chusing, as Doctor *Bacon* saith, to congeal the thing that Nature begun her first operations about, by a proportionate mixtion, and union of pure living *Mercury*, with a like quantity of Sulphur, into one Mass: Oh holy words! wherein this good Anglian, or rather Angel, clearly depinged that one and true matter, whereof all Philosophers have writ Volumes under divers figures and Enigmatical Fables; not because they would malitiously hide it, but keep the priviledge of this knowledge for learned and pious Men, who by continual study and laborious experience finde and adorn it.

But lest I should move some masters to suspect that I alledge this place ignorantly, and understand it improperly; I would have them know, that by that matter which *Bacon* so ingeniously represents, I mean the universal Spirit whereof I treat ; and likewise, that I put a difference between the Father and the Son ; or the Genitor, and him that's Generated ; or the Producer, and him that is Produced : neither need I blush to say, that I know the one as well as the other : For the Philosopher here would have such enquire after the confection of the Philosophers Stone, to seek the principle of Minerals ; and he paints out the first matter of Metals, prepared, compounded, and specified by Nature.

But I treat of first matter not yet specified, which may be properly called the first matter of this first matter of Metals, or the most general Genus, so much celebrated by *Raymundus Lullius* ; but I used this sentence for example and authorities sake, yet so as no absurdity lurks therein : for the universal Spirit is the common Parent of Mercury and Sulphur, contained and proportionated by Nature, in this one Philosophical subject.

But I would have the curious Artist consider two things : first, that by subtile imagination he chuse an enlivening

Nature, apt for the conservation of all Bodies; the other, that he chuse a thing which of it self can enliven, and regenerates. Yet I would not have him to chuse two different & separated Matters, the one Agent, and the other Patient; but onely one, that may at once be of vertue to enliven, and to be enlivened.

As to active Vivification, I have said enough; but as to the Passive, I say, That every Principle hath its Original from it self: For if it should have it elsewhere, it were no Principle; and while it gives being to others, it must necessarily, whilst it generates them, draw from it self restauration, and perpetual plenitude: wherefore it is in continual action and motion towards vivification, whereby its destruction is hindred; for it will never forsake it self: and it hath motion in and from it self; which *Macrobius* also disputed in his Comment upon *Scipio's* dream, discoursing on the soul of man; though I think his discourse may be better apted to the Soul or Spirit of the world, which is my subject.

I will therefore borrow this from his Arguments: Whatsoever is moved of it self, is the beginning of motion, and lives continually; and that that lives continually cannot be enlivened but from it self, it is therefore vivificable: but the Spirit of the world is such, because it hath its seat in the Earth, to convert it self into Earth, wherein, as *Hermes* rightly, all it vertues, actions and qualities remain entire: it followes also, seeing it is vital, that it reassumes life, restoring it self by its own proper power: we find the same also in this universal Mercury, which is always nourished, and restored in its Myne; so that if by any means it be extracted, it always grows to the same form whereof it was before; and wheresoever it is cast, there will be plenty of it always after.

Not as if it were generated of the Earth, but in the Earth, through whose parts it creeps, extending it self continually by multiplication and vegetation; which the Ancients denote by the Serpent, which *Moses* says creeps on the Earth, & feeds it self on the dust thereof: and this caused the *Cabalists* to call him the prince of Sepulchres, because he devours and consumes the Bodies there interred; not that dead Bodies, or the earth, are his aliments, but only the seats where he is fed and nourished. This is the place where he is moved, turned, and twines without ceasing; whereof *Medea* admonishes *Jason*, where she says (in *Epist. Heroid. Ovidij*)

Pervigil ecce draco squamis crepitantibus horrens,
Sibilat & torto pectore verrit humum.

Lo here the Dragon with his horrid Scales,
Doth watch, and hiss, and plow the very dales.

Which *Rythmes* a French Author thus expresses.

Voy le dragon veillant, de fureur forcene
Qui d'escaile crugaten a le corps en tourne,
Dont le gosier sifflant fumee & feu deserre
Et qui par replis lors va baliant la terre
De sa large poitrine en la poudre imprimant
Les sineux siallons qu'il trace incessament.

Behold the scaly swelling Dragon lurking,
Who always listens with a watchful ear (here
Who knits his brows, and never shuts his eyes
But him that sees his cust tongues, teeth a-frights

With horror; whose wide throat emits such
                                   (flames,
As do infect the Air with blackest fumes:
Behold his many twinings, which he deep
Impresses in the earth, whilst he doth creep,
And plough the ground with his broad brest,
                                   (whilst he
Returns in the same tract continually.

I adduce these two Considerations, not onely to shew how *Mercury* must be sought, but also to confirme, that that which is fixable in it, is nothing else but the enlivening essence, which fixed in due manner, perpetuates and keeps life in all things it enters; by its purity expelling Excrements, and by its perfection, perfecting imperfect things. The end of Fixation, both natural and artificial, is Perpetuation and Conservation, which are effected by the Mediation of that Tincture which *Mercury* acquires by this Fixation: for that Tincture is Life, and Life is nothing else but that which opens and colours the Body with such a Tincture, as shews it to be vital, and perishes with its death.

Nature therefore colours Blood,

wherein life consists, with a red Tincture: and when the Blood is clearer, and more lively red, the Body is more sound, fair and vigorous: as on the contrary, when the Blood is dense, black, adust with choler, or changed into a false colour, the Body is pained and sick within, and by discoloration gives Testimony thereof without.

We may observe the same in Vegetables, whose lively vigour consists in greenness; which being changed, we say it is turning or declining towards death. The perfection also, or imperfection of Metals, is discernable by their colours. Gold is indued with a magnetical vertue, which by the splendent fulgor of its tincture, draws man's earth after it: in which Nature spends all her forces, but leaves the victory to Arts industry, which by graduation to the haight, which it adds to its natural splendor, makes it far more fulgent; insomuch that it's called the Terrestrial Sun.

An Artist then may exalt the golden colour to the height of obscure redness; by which augmentation, imperfect

Metals in a certain degree, may, by projection of this artificial Tincture, be brought to the height of perfection; so that we see this golden colour introduced by Nature into this Metal, is onely the way to that redness, wherein the completion of perfect Vertue lies: for which cause, this Metal, though far excelling others, can communicate no perfection nor conservation to humane Bodies, as a thousand Jugglers and Sluggards in Physicks, promise by their Sophistical Fusions, and Phantastical Confections.

But if more curious Artists work upon this subject, they may make it acquire such a degree of inseparable redness, that by the excess of its heat, it shall work miracles, and yet it shall consume nothing but superfluities, and shall conserve and multiply the substance of Bodies; though Philosophers say, That its heat as much exceeds our common fires, as common fires do innate heat in Animals.

Paracelsus in his Treatise of Tinctures, extols that highly which is extracted out of Gold by Spirit of

Wine, and attributes many singular Vertues to it, as also to that that's made of Antimony and Coral; before which, he yet seems to prefer the Tincture of *Mercury*, which he says may, by perfect Fixation, be brought wholly to a Tincture, so that it will penetrate Bodies, because of its most subtile purity: where I think, he means not that vulgar, but Philosophical *Mercury*, wherein Art, perfecting Nature, hath wrought these two effects, to wit, perfect Tincture, and compleat Fixation.

Tincture then in proper locution, is the pure substance of things; and Body is nothing but an Excrement: which is also manifest, in that Bodies, after the Separation of their Tincture, are useless, without vertue, and corruptible; no otherwise then a carcase without life, colour or motion.

Tincture may then be called the scope of Fixation, it attaining by its permanency in fire, a conservative faculty in those Bodies to which it is applied. But the manner of attaining this degree of Fixation, in which the Completion of the whole work con-fists, is no other, then that fugitive and light things be prudently kept in the Fire, that they may be brought into assuefaction with it, that they may endure most violent heat. And for this cause, good Authors commend Patience to their Disciples, as proceeding from God; but Precipitancy, as from the Devil.

Take this for an infallible Rule, That unless Calcination go before, nothing can be fixed; and that this should be done, by conjoyning the fixable Spirit with something of a convenient Nature, that may retain it in the Fire of Calcination, that by this means it may accustome it to sustain heat by little, till it can endure the ultimate augmentation of Fire, which infers Fixation. And the Reason why we must proceed with such discretion, is, because, if we should too readily precipitate this operation, the special Spirituality, which is the Mother of this Tincture, would flie away, and leave the Body without any impression of the tingent Vertue; so that a new Spirit must of necessity be given to this dead Head, before the desired colour can be introduced, which is one of the Secrets of this Chymical Art: for it is the Spirit, and no other thing, that colours by mediation of Fire; and this Tincture compleated and exalted in our *Mercury*, should be elevated to the height of perfection; that, as *Hermes* speaks, it may ascend into Heaven; and when it hath sustained all mortal torments, receive a new life; that is, after it hath passed the darksome straits of Putrefaction, it may be elevated to a Resurrection, by the Ablation of all mortiferous and corruptive Accidents. And thus it will attain the height of perfection, which is effected by Separation of Earth from Fire, the subtile from the spiss; and by Fixation of its purged part by gradual heat.

But that I may speak without ambages and doubts, This ascent into Heaven (which is the Sublimation and Exaltation of its parts to perfection) cannot be effected, unless Separation and Purification go before, and give place to Fixation, as to the scope and ultimate end of Art.

And here note, That this is done for

two ends: one is, That the Tincture may be perpetuated; the other, That the volatile and combustile Sulphur of the *Mercury*, may be separated and extracted; which cannot be effected, but by the long and continual action of Fire: and this Fire must be regular, lest violent precipitation in the beginning, make the pure Spirit of *Mercury*, not yet fixed, to ascend: which *Comes Trevisanus* hints at, saying, That Writers differ about the structure of the Fire, though they all aim at one and the same scope, to wit, That it should be so made, that the fugitive Spirit should flie away, before the persequent suffered any thing from the Fire; that is, that the spiritual part should not leave the corporal, through the ardour of the Fire, which should fix it by the action of common Fire, discreetly applied in its several degrees; wherein the whole of this Art consists.

But some may say, If Fixation in the Fire communicate Permanency to this penetrating subtilty, how shall it afterwards be sublimated?

Let such take waxen Wings, and they

shall see, that they will have a minde to flie from this Prison, the Earth: But they must minde, lest ascending too high, the Sun melt their Wings, and burn their Feathers, and so precipitate them into the Sea. But let them imitate wise *Dædalus*, who held the mean betwixt two extreams, because, if he had flown too low, the Water would have loaden his Wings; if too high, the Sun would have melted his Wax. It was the impatience and blinde desire *Icarus* had of overcoming *Dædalus*, that wrought his ruine.

And whence came *Phaëton's* pernicious ruine, when he would govern the Sun, but that he thought himself more apt for this work, then his Father? who admonished him thus:

*Hac sit iter manefesta vota vestigia cernes
utq; ferant æquos, & Cælum & Terra calores
Nec preme, nec summum molire per æthera cur-
Altius egressus Cælestia signa cremabis, (rum
Inferius Terras, medio tutissimus ibis.*

Run where thou seest the marks that I have
(made
With these same wheels; and that thou maist
(evade
The danger of burning Heaven, Sea, or
(Earth,
Flie not too high, nor yet stoope down be-
(neath
This tract; for if too high you soar, you'll
(fire
The Heavens; if too low, the Earth you'll
(move with ire ;
Keep then the mean, as safest, in your gyre.

But the recitation of these Ovidian Verses, are not enough, though they be most true, according to the Opinion of the Ancients; I will rather explicate their intention folded up in the Coat of their Fables, seeing those onely serve to the expert in this Art.

Let the curious then know, That when *Hermes* says this thing must ascend into Heaven, and again descend to the Earth, and acquire the vertue of both; he means not that the matter should be sublimated to the top of the Vessel, but onely that it is necessary, that after perfect fixation, some spiritual portion be applied to it, whereby it may be dissolved, and become altogether spiritual; leaving its Terrestrial consistence, and assuming an aery Nature, which is the Philosophers Heaven; and when it hath reached this simplicity, be again coagulated and reduced to Earth, by a new coction effected by the same degree of heat, till the Body so imbrace the Spirit, that they become one incorporated, and by this means acquire a Celestial subtilty, and a Terrestrial fixation.

And that we may always note Natures ways; if this *Icarus* cannot totally elevate himself, he must resartiate his wings, by putting to new wax, new feathers; that is, by reiterated dissolutions, which the Masters of this Art so oft repeat, that they seem importunate to all such as understand not the consequence of this repetition; which yet is onely, that things might be better united, being mixed by their least particles; which none can effect without the purification both of the Body and Spirit, keeping the Spirit volatile from all Terrestrial impurities, and the Body from all internal Dregs, during this fixation: These things then ascend into heaven by Dissolutions, and descend to the Earth by congelations.

This Body then glorified, will ascend into Heaven upon the wings of its Spirit, and in the same perfection again descend to the Earth, to separate good from evil, and preserve the one but destroy the other: that is, what Bodies soever it enters, it ejects their impurities, and conserves their purer substance; for reiterated solutions and fixations, gave it power to enter Bodies. This *Hermaphroditical yong man, and his delicate Salmacis*, must be washed in a Fountain, that they may embrace each other; and *Salmacis* burning in love, say, When shall the time come, that this fair Yong man shall never be separated from me, nor I from him, and that our mutual loves may perpetuate their conjunction in felicity, that so these two Bodies may have one heart and one face? and then we must take care, that the *Island Delos* remain immoveable, and that *Apollo* and *Diana*, whom *Latona* there brought forth, may be both stayed in that place. This fable denotes nothing, but that the dissolved matter, wherein the Philosophers sun and moon is contained, should again be congealed and fixed.

I would not have my Reader imagine, that he shall in this Book finde the rich Mines of *Peru*, to satiate his avarice, and make him rich withal; though I have sufficiently demonstrated in several places, to men that have their eyes in their heads, that the way to these riches is not unknown to me. Yet I will not easily be perswaded to undertake so long a journey, for certain reasons, not unlike those that held *Trevisanus* two years from this enterprise, after he had got the perfect knowledge of this Magistery. Onely I have in my minde determined to ratifie that most precious confection, or rather inestimable treasure, which benign Nature gives for the sustentation and prolongation of mans life, who from God received priviledge to be mans Protector; which I undertake, out of an honourable desire, that I may by my industry advance the publike good, when the favorable Star of experience hath led me to my secure Port, which I would willingly impart to curious Men: For I have so successfully elaborated this Universal Spirit, that with a small quantity, I have restored above a hundred men, consumed with different Diseases, to sanity: and doubtless many excellent wits would have penetrated deeper into this obscure and devious Wood, but that seeing it filled with horrid Monsters, they have been so confounded, as to leave this perillous Path, and forsake their enterprise.

As *Poliphilus* hath by a most ingenious pencil expressed; whose generous and undaunted courage, sleighting those common terrors, hath effected so much, that both the sides of this black, dark Word, lay open to light; and by whose manaduction you may, notwithstanding all obstacles, arrive safe and and sound at the delicious and grateful habitation of Lady *Polia*, shut up in the rich Temple of *Vesta*.

This I can assert for indubitable, that

the way he held is open to all; but all have not that *Ariadne's Clew* that he had, to extricate himself out of this labyrinth; neither is every one a *Theseus*, that he can overcome the *Minotaure*.

It is certain that Nature, like a loving Mother, proposes and offers this precious Treasure of life to all; and that God our Universal Father, keeps open the Gate of this fatall Cavern, for the commodity of all men: for the descent to Hell is easie; but then again, to ascend to Heaven, *hic labor, hoc opus est*: Stay coach-man, here's a straw.

First then, that splendent Branch, Dedicated to the infernal *Juno*, must be found out, of which *Virgil* writes.

Accipe que peragenda prius, latet arbore opacæ
Aureus & foliis & lento vim ine ramus
Junoni inferne dictus sacer, hunc tegit omnis
Lucus, & obscuris claudunt convallibus umbræ,
Sed non ante datur tellures operta subire,
Auricomos quam quis decerpserit arbore fætus.
Hoc sibi pulchra suum ferri Proserpina munus
Instituit, primo accurso non deficit alter
Aureus & simili frondescit virga metallo.
Ergo alte vestiga oculis, & rite repertum
Carpe manu; namq; ille volens facilisq; sequetur
Si te fata vocant, aliter non viribus ullis
Vincere, nec duro poteris convellere ferro.

First see what you must do: the golden tree
Which to th'infernal Juno we decree,
Lies hid under a grove in a deep vale;
And now you may not pass within the pale
Of th' Earths deep Cabinet, until you have
Snatch'd from this tree her off-spring, which
(to grave
Proserpina you must present: the tree
That gave this fruit, will not long fruitless
(be:
For loe, another golden off-spring follows;
(then
Fix your eyes here, and in your hand retain
What you do finde: if th' fates do favour
(now,
You vanquish; but if not, no force will do.

If therefore Nature be follicitous in hiding these things, lest they should be indifferently prostrated to all, *or Hogs get to the honey-pots*; no wonder if Ancient and Modern Philosophers, have invented so many ænigmatical Figures, and hidden Fables, to cover and cloath this Science with; For they know well enough, that ceremonious Nature, would never have hid her self under so many different Forms and Species, but have appeared naked, but that her Venerable Secrets would thereby incur that contempt, which alway accompanies common things.

For which cause I also in this Treatise use the same Solennity and Taciturnity, left I should undergo the same peril that he did, who published the Mysteries of the Elysian Goddesses, which should have been kept Secret and Chaste, and not like common Harlots exposed to publike abuse: And whether I have spoken conveniently to my own purpose, or no, such as have made progress in this search, may better judge: for experience is the best Dame.

In the mean time, I hope none will give a hard interpretation, or pass a bad sentence on me, because I have compared Natural and Chymical operations, with Divine and Christian Mysteries, wherewith they have some conformity: For by this application, I have no way profaned them, but rather Celebrated their Excellences, and pointed at that great care and testimony of our Creatour, in ratifying the safety both of Body and Soul together.

Which reason moved one Author, so that he wrote, that the true Chymistry, which *Paracelsus* calls *Spagiry*, follows the Gospel foot by foot; because by the help of this and fire, Nature exerts her most potent faculties; which Antient Philosophers, as *Brachmannus*, *Gymnosophisters*, and all the Egyptians, insinuate in their *Theology*.

For all the Magick of Paganism, and all the fables of Poets, signifie no more then what this Book Treats of; which the most Learned and subtile *Bracescus* diligently examined; though envious *Toladanus* writes the contrary, after he was deceived in an experiment in this Art, which he thought he had wrested from him by his importunity, believing that the spume of Iron was the Philosophers Mercury, because he asserted that it was extracted out of a vile Matter, of small price, which was cast into the streets; not observing, that the discreet Masters of this Art, cloath their Matter with strange Vestments, calling them by all the names of Metals, indiscriminately, and yet without the least fraud: for he that knows this matter, knows also, that it contains the seven Metals in it.

And I would gladly know, whether they think *Cosmopolita* meant of the vulgar Steel, when he sayes, That *Neptune* shewed him under one Rock, two Mynes, the one of Gold, the other of Steel.

The Man was too plain, to harbour such frivolous conceits: but he named his matter thus, because of the conformity with polished Steel. And *Bracescus* had played the part of a Fool, and not a Philosopher, if he had in one moment opened the whole myftery of his Secret, in the acquiring whereof he had doubtlefsly spent two thirds of his life.

But that I may add something to the explication of these Mythologies, do we not plainly see, That that ancient *Demogorgon*, the Parent of all the gods, or rather of all the Members of the World, which, they say, inhabits the Centre of the Earth, covered with a green and ferruginous Coat, nourishing all kindes of Animals, is nothing else but the universal Spirit, which by God's command produced the Heaven out of the Chaos, with the Elements and all things therein, which it always hitherto sustains and enlivens? for it truely took up its Habitation in the middle of the Earth, as we have declared in the beginning of this Book, where it sits as it were in its Throne; and thence, as from the Heart of some great Body, and Seat of Life universal, animates and nourishes all things: and that green and ferruginous Coat, is nothing but the surface of the Earth, which is of a blackish and Iron-colour, decked and variegated with divers coloured Herbs and Flowers.

Virgil, perfectly skilled in all these myſtical Secrets, called this Spirit or Soul of the World, *Jupiter*; whom his Paſtorer *Damæta* invokes, in his third Eclogue, becauſe, as he ſaith, all things are full of him. And *Pan*, that god of the Woods, worſhipped by Shepherds, is taken for the ſame: for beſides that this word *Pan* ſignifies *All*, he is alſo made the God of the Woods, because the Greeks, for the Governour thereof, worshipped the Chaos, which they otherwise call *Hyle*, that is, a Wood. *Orpheus* bespeaks *Pan* thus:

Pan le fort le subtile, l'entier, l'universel,
Tout air, tout eau, tout terre, & tout feu immortel,
Qui sieds avec le temps dedans un throne mesme,
Au regne inferieur, au moyen, au supreme.
Contevant, engendrant, produisant, gardant tout,
Principe en tout, de tout, qui de tout viens a bout,
Germe du feu, de l'air, de la terre, de l'onde,
Grand esprit avivant tous les membres du monde,
Qui vas du tout en tout les natures changeant
Pour ame universelle en tous corps te logeant
Ausquels tu donne estre, & mourirment & vie
Promant par mille effects ta puissance infinit.

Pan strong and subtile, great and general,
That art both Fire, Air, Earth, Water, and
(*all*;
That raign'st alwayes, and over every
(*thing*;
Getting, conceiving, bearing, and keeping
What'ere beginning had, or have an end;
That Art, both Tree and Fruit, both Foe
(*and Friend*;
The Spirit that doeth binde the worlds parts,
That penetrates all Natures, and imparts

Both life and motion, power to act and will,
And that dost with thy vertue all things fill.

Saturn the Son of Heaven and *Vesta*, which is the Earth, and Husband to *Ops* his Sister, (which is that co-adjuvant and conservant power of all things) represent *Demogorgon*; and the Infants that he first devours, and then vomits up again, are they not Bodies, to whom he gives being, and at last reduces them to himself, whence new ones alwayes issue, that by his perpetual vicissitude, the order established in the beginning, may be preserved to the end?

He is sometimes painted with grey and sordid Hair, his Head covered, with a Sickle in his Hand; and for his Symbole, having a Serpent circularly grated, and holding its Tail in its Mouth: And sure he is old enough, being the Grandfather of all: His Beard and Hairs are white, alwayes growing; which are the things that alwayes germinate: He is sordid, and ill disposed of himself, because of the Terrestrial impurity which adheres to it, by reason of its Sulphureous and Adustive Corruption: His Head is covered, that is, the principle of perfection is closed in the cover of impurity, which makes it known to few: His Sickle is his Penetration & Mordacity, wherewith he penetrates & devours all things: The Serpents biting its tail, is his regenerating nature, wherby he repairs himself; as is storied of the Phœnix, by wch name he is also sometimes called: so that he is always conversant in a circular and indeficient encrease.

But methinks I hear some say, That I miss of the intention of Authors in their fabulous description of *Saturn*; for they mean by *Saturn*, Lead; which is of all Metals the Seniour: which also devours all others by its crudity, because it is full of Salt: for Mordacity arises from Salt, as the Refiners of Metals finde in their probations: for it there vomits out the Gold and Silver, which it deglutiated, but could not consume, because in its decoction they received Fixation, so as to resist the weak heat of its Ventricle.

I do not altogether reject this sense,

because it is in some places conformable to the other; and in that it is in all things concordant with what I have given, I hope none will tell me mine is false.

Maia represented the Earth; so called, because out of her, as a Mother, the universal Spirit, or *Mercury*, had its originé; and from the pure invisible Seed of *Jupiter*, which is the Air: for it proceeds really out of these, as the learned *Cosmopolita* sayes in his precious Treatises.

Mercury is commonly painted with Wings, that they may shew how she is of a volatile Nature: his Head is covered with a Hat, for the same Reasons that I before gave for *Saturn*'s Head: He carries a caduceous and fatal Rod, encircled with Serpents, both to signifie and denote his regenerative faculty, as those Reasons also I gave to *Saturn*'s Serpent. By this Rod, he opens Heaven and Earth, and gives life and death: and that Rod represents powerful Nature: for by ascending into Heaven, and descending to the Earth, he acquires the vertues both of superiour and inferiour things. By this same power, he draws Souls out of Hell, makes all Eyes yield to Sleep, as *Virgil* writes of him.

He is by some also called, Poyson and a Theriack, or Life and Death, according to his use and dosis: for Life consists in due temperament and justice; Death, in excess.

Many other such mysteries are contained in this Heathenish Theology, which have no other scope then that I aim at: all which, if I should here explain, my Treatise would prove a Volume. But I will not defatigate my Reader with frequent repetitions of one and the same thing.

These may suffice to shew, That all Mythological Commentaries, such as these, with their Allegorical and Historical Sense, gave no occasion to Poetical Fictions, as if any Truth were in them: but such things tend chiefly to shew the admirable operations of Chymistry, as amongst others, the Story of *Jason* and *Medea*, explicated by *Chry-*

Sogonus Polydorus. To which Explication I will add, That this Name *Medea*, that signifies Cogitation, Meditation or Investigation, is derived from a word that denotes a Principle, Origine, or Reason: for all Meditation hath doubtless some Principle or Reason for its Foundation.

This *Medea* taught *Jason*, (that is, an Inquirer) two things, wherein all Philosophy consists: first, The Acquisition of the Golden, that is, the Art of changing Metals and Minerals. And secondly, The Restauration of Bodies weakned with Diseases, readily and perfectly curing them, and bringing them from old Age to Youth and Vigour, by this sole and universal Medicine; expelling all corrupt and corrumpent humours and superfluities from Bodies, which would else bring them to their end.

Jason endeavoured and perfected these miraculous effects, by the observation of *Medea's* counsel, after long and laborious Navigation, obnoxious to many perils, both in sailing, in killing the Dragon, and taming the Bulls.

This Navigation is the laborious Inquisition, and dubious Experience of things, wherein many are exercised all their life long, and cannot arrive at the Port of Nature.

The monstrous Bulls which he was to tame and subjugate, are the Furnaces wherein the operations are made, which do not ill represent a Buls Head; and breath out Fire at their Mouth and Eyes, as the Fable hath it: for they must needs have some Transpirations, else the Fire would be extinguished, and the degree of its heat not regulated: for unless some experienced Man regulate the Fire, many things would fall out wrong in the operations, and delude the Operators hopes, as I have experienced; for of nine Vessels, which I reposed in a Furnace, to give it a due degree of heat, I lost eight, and one I conserved; by means whereof, I obtained the said Experiments in curing Diseases.

The watchful Dragon, is that universal *Mercury* which *Cadmus* had learned to kill, that is, fix the *Campus Mortis* wherein the teeth of the Martial Serpent should be sown, is nothing but a vessel wherein Souldiers, armed with Spears, are elevated.

But this Vessel should not be a Glass-Alembick, as *Polydorus* thinks, but made in form of a Cooperculum or Covert, narrow below, and capacious above; made of good Earth, well cocted, and not of Iron or Glass. In whose bottom, *Mars* his Field must be elevated, rigid with Lances and Spears, representing souldiers provoked and fighting.

And this is onely an ingenious Fiction of the Poet, to make the thing somewhat admirable to the vulgar; which yet is so plain and familiar to us, that if I should name what it is, I should make my self ridiculous.

And when *Jason* had finished his work, he must needs make the watching Dragon, that kept the Golden Fleece, sleep, that he might no more eructate Fire and Fume; which he did, by suffocating him in the Stygian Waters, that is, by dissolving and fixing him with his Spirit. And then *Jason* had nothing to do for the possession of the Golden Fleece, and restoring his old Father *Æson* to Youth again, but onely one business that *Medea* taught him, that is, the fermentation and conjunction of solar butter, with the paste of prepared Mercury, which is not of it self fit for two such excellent actions, being onely Earth, wherein pure ferment produced by Nature, and promoted to perfection, should be sown.

When he had finished this last work, he saw himself possessor of his twofold Treasury, which he gloriously brought back to the place of his Nativity; by benefit whereof he became very rich, and reduced his aged Father to sanity, by removing his languishments.

But I will now leave *Jason* with his *Medea*, to the enjoyment of their felicity, and onely adde this, That by the Dragon that watched continually, and vomited fire out of his throat, nothing can be more conveniently denoted, then our Universal Spirit or Mercury, which is the most vivacious and inflammable thing in the world; wherefore it is called burning water, or *aqua vitæ*, because, as *Bracescus* notes, it always burns before its coagulation; and it is *aqua vitæ*, because it vivifies all things: And if one should look on its superficies, who would think it fixed, and inconsumeable, being so easily inflammable, or vanishing at the least touch of fire? or who would think, that there were any conservative vertue in its Centre? seeing it looks as if it were full of mortal poyson, rather then life?

But as God set a Cherubim, with a flaming Sword, to keep the tree of Life: So Nature set this ignivomous Dragon, in the door of the Garden, to keep the Tree of golden Apples, that is, the knowledge of her hidden Secrets, which our prudent Ancestours would not deliver in writing, but onely by word of mouth, to such as they thought worthy of such knowledge: And this is the cause, why those great and admirable Sciences, have in progress of time vanished, and are accounted as Fables and Tales.

Which thing *Esdras* considered: for foreknowing that the Israelites should suffer banishments, flights and captivity, he feared left the secret Mysteries of the Scripture should perish; because without the benefit of writing, mens memory might easily fail.

He therefore Congregated all the Elders, in number LXX, who with himself wrote all these things in as many Books, as himself attests, saying, That after forty, days the Lord spake and said, *Publish those things which thou hast written*, that all may read them: and keep those later LXX Books, that thou maist deliver them to the Wisemen of thy people: for herein is the vein of Understanding, and the Fountain of Wisdom, and the Flood of Knowledge; and so he did.

Picus Mirandulanus, the Phenix of his time for Learning and Knowledge, speaks of those Books with great reverence, in these or the like words.

These are LXX Cabalistical Books, wherein *Esdras* said plainly, That the

Fountain of all Understanding and Knowledge, was contained, that is, the inestimable *Theology*, concerning the Supream Deity, the Fountain of Wisdom, and the entire Metaphysicks of intelligences; the stream of Knowledge, that is, the firmest natural Philosophy.

When these Books had been long kept secret, they were by the command of *Xystus* the fourth, turned first into Latine, for the benefit of Religion; but this good Work was interrupted by his death: Yet they are had in such veneration amongst the Jews, that no one under forty years of age, may touch them: And this is most admirable, that in those Cabalistical Doctrines, some Heads of Christianism should be contained. Thus *Mirandula*.

And now having, I hope, omitted nothing, which might serve to the interpretation of *Hermes* his Table, or obscure Philosophical Cabala; I shall betake my self to the Port of this unsearchable Ocean, and dry my wet cloathes under the Sunshine of your

Favours, wishing well to all; and upon good grounds, judging that the Good, Honor, and Glory of the World, is the true Philosophy.

FINIS.

www.ingramcontent.com/pod-product-compliance
Lightning Source LLC
Chambersburg PA
CBHW060833170526
45158CB00001B/158